A Different Hunger

A. Sivanandan

A Different Hunger
Writings on Black Resistance

'Theirs is a different hunger — a hunger to retain the freedom, the life-style, the dignity which they have carved out from the stone of their lives.'

First published in 1982 by Pluto Press
345 Archway Road, London, N6 5AA
and 141 Old Bedford Road,
Concord, MA 01742, USA

Reprinted 1987, 1991

ISBN 9780861043712
ISBN 0861043715

Printed and bound by CPI Group (UK) Ltd,
Croydon, CR0 4YY

Contents

Introduction

by Stuart Hall

This collection of Sivanandan's writings is long overdue. The majority of the essays were first published in *Race & Class*, the journal (published by the Institute of Race Relations) which he founded, edits and has sustained through thick and (mainly) thin times. For those who do not know, it is worth saying that he is one of the handful of key black intellectuals who has actively sustained the black struggle in Britain over more than two decades: partly in his writing and educational work; partly — and less visibly — in political interventions of a strategic kind; partly by his defence of the Institute as a base for active political work; and partly by his own considerable personal gifts and qualities. The so-called 'unity of theory and practice' has been frequently invoked on the left, but remains elusive. These essays represent something as closely approximating to that ideal as one is likely to find anywhere in recent British writing on the themes of black struggle.

The essays deserve to be better known. Many of them will be already familiar to those engaged in the politics of black resistance; though it is good to have them at last between single covers. They have acquired, indeed, a remarkable 'underground reputation': thumbed over, read and reread, argued about and debated wherever these issues are taken seriously. They have also, frequently, been plagiarized — which is its own kind of recognition: usually without acknowledgement, which seems to be the fate, not only of Siva's work, but of the Institute — the first telephone number which journalists and other seekers after truth can find when in search of 'the black connection', and the name they find most difficult to recall when the credits and acknowledgements begin to roll.

Even amongst white activists who regard themselves as fully paid-up members of the anti-racist struggle, they are much less familiar than they deserve.

I doubt whether these are matters that keep Sivanandan awake at night. Indeed, I suspect they must be numbered amongst the many rich ironies of political struggle which he so deeply relishes and from which he has extracted so much humour over the years. This capacity to take some pleasure from the richly contradictory nature of black struggle in a white society is a quality which has infuriated numerous white audiences, and is not always well understood even amongst his friends and would-be supporters. The black cause, as we know, is a serious matter: hence black intellectuals are expected to be serious, noble and dignified at all times — above all, *simple* in their indignation. Sivanandan has, however, systematically refused to be 'simple' in this way. I have always regarded this complexity — in the best sense, his 'doubleness' — as a political, as well as a personal, strategy. Surrounded as he has been by the ambiguities, duplicities, the multiple masks of his so-called 'host society', Sivanandan has actively cultivated a number of appropriate *personae*. One face — extremely polite, especially when pointing out the contradictory toils in which the speaker is currently entwining himself: and full of a rich but cunning show of innocence — is sometimes turned to allies and enemies alike, 'above ground'. Another — principled, militant, intransigent in opposition: yet gentle in personal relationship — is reserved for comrades and friends with whom he has become linked and bound in struggle, 'below ground'. His enemies would be well advised not to mistake the one for the other.

The quality, however, which most distinguishes these essays is simply his capacity to go directly for the seminal issue, and to give that issue an original formulation. The essay on blacks and the state, for example, inserted the black question into the very centre of a growing and wide-ranging concern — novel at the time — with shifts in the strategy of the state. The formulation of the issues involved in the qualitative shift from 'controls' to 'induced repatriation' was so clarifying that it became the adopted wisdom overnight. It formed a watershed in the thinking about the structural position of blacks in Britain. The connections between blacks workers, the 'new technology' and the new international division of labour were powerfully posed in what for years remained as the only political discussion readily at hand. The analysis of such key struggles as Grunwick and the riots of Summer 1981 has become something of a test-case for political analysts of the left. Siva's accounts are amongst the few which deserve to be preserved and which stand re-reading years after. These and other pieces in the collection are *focussed* by the imperatives of struggle; they bring ideas to bear on clarifying its perspectives; and they flow back into the strategies of struggle. Few writers can claim to have approached so clearly the organic connection between analysis and practice, which is the hallmark of the political intellectual.

Those involved in or committed to these struggles will, therefore, have much to gain from reading or re-reading these essays. But — historical memories being so short — it is worth reminding them of the story of how the base for this kind of work was laid. Behind these essays lies the history of the Institute itself: the focal point for 'race relations' research in the early days; then, like CARD (the Campaign Against Racial Discrimination), polarized and fractured by the growing politicization of black struggle in the 1960s; the critical moment when the 'race relations industry' was first identified, and its project analysed — these are crucial moments, at each of which Sivanandan played a critical role. Few know the story of how he simply hijacked the Institute from under their very noses; took the material resources (books, journals, pamphlets, filing cards and connections) which he has helped, painfully, to accumulate, packed them up, and walked out with them, as it were, under his arm; transferring them to a less salubrious and less respectable part of town, bearing the official title (to the establishment's intense annoyance) with him, to a base where the brothers and sisters had a far better idea what to do with them than those he had left behind. Few librarians have achieved so striking — and brazen — an appropriation/expropriation of the tools and materials of their trade! It was an inspired act of piracy which should illuminate our path and deserves to be regularly celebrated: revealing an impeccable sense of the moment, and an instinctive set of priorities; above all, thoroughly characteristic.

These are papers and essays 'from the front line'. I greatly envy those who are about to encounter them for the first time.

I would like to thank the Transnational Institute, Washington, for their support during the writing of some of these pieces.

A. Sivanandan

Part One:

Asian and Afro-Caribbean Struggles in Britain

From Resistance to Rebellion

On 25 June 1940 Udham Singh was hanged. At a meeting of the Royal Asiatic Society and the East India Association at Caxton Hall, London, he had shot dead Sir Michael O'Dwyer, who (as the Lieutenant Governor of the Punjab) had presided over the massacre of unarmed peasants and workers at Jallianwala Bagh, Amritsar, in 1919. Udham was a skilled electrician, an active trade unionist and a delegate to the local trades council, and, in 1938, had initiated the setting up of the first Indian Workers' Association, in Coventry.

In October 1945 at Chorlton Town Hall in Manchester the fifth Pan-African Congress, breaking with its earlier reformism, pledged itself to fight for the 'absolute and complete independence' of the colonies and an end to imperialism, if need be through Gandhian methods of passive resistance. Among the delegates then resident in Britain were Kwame Nkrumah, Jomo Kenyatta, George Padmore, Wallace-Johnson, C.L.R. James and Ras Makonnen. W.E.B. DuBois, who had founded the Pan-African Congress in America in 1917, presided.

In September 1975 three young West Indians held up a Knightsbridge restaurant for the money that would help set up proper schools for the black community, finance black political groups and assist the liberation struggles in Africa.

Of such strands have black struggles in Britain been woven. But their pattern was set on the loom of British racism.

In the early period of post-war reconstruction, when Britain, like all European powers, was desperate for labour, racialism operated on a free market basis — adjusting itself to the ordinary laws of supply and demand. So that in the sphere of employment, where too many jobs were seeking too few workers — as the state itself had acknowledged in the Nationality Act of 1948 — racialism did not debar black people from work per se. It operated instead to deskill them, to keep their wages down and to segregate them in the dirty, ill-paid jobs that white workers did not want — not on the basis of an avowed racialism but in the habit of an acceptable exploitation. In the sphere of housing, where too many people were seeking too few houses, racialism operated more directly to keep blacks out of the housing market and to herd them into bed-sitters in decaying inner city areas. And here the racialism was more overt and sanctioned by society. 'For the selection of tenants', wrote Ruth Glass scathingly,

> is regarded as being subject solely to the personal discretion of the landlord. It is understood that it is his privilege to bar Negroes, Sikhs, Jews, foreigners in general, cockneys, socialists, dogs or any other species which he wants to keep away. The recruitment of workers, however, in both state and private enterprises is a question of public policy — determined explicitly or implicitly by agreements between trade unions, employers' associations and government. As a landlord, Mr Smith can practise discrimination openly; as an employer, he must at least disguise it. In the sphere of housing, tolerance is a matter of private initiative; in the sphere of employment, it is in some respects 'nationalised'.[1]

That same racialism operated under the twee name of colour bar in the pubs and clubs and bars and dance-halls to keep black people out. In schooling there were too few black children to cause a problem: the immigrants, predominantly male and single, had not come to settle. The message that was generally percolating through to the children of the mother country was that it was their labour that was wanted, not their presence. Racialism, it would appear, could reconcile that contradiction on its own — without state interference, laissez-faire, drawing on the traditions of Britain's slave and colonial centuries.

The black response was halting at first. Both Afro-Caribbeans and Asians, each in their own way, found it difficult to come to terms with such primitive prejudice and to deal with such fine hypocrisy. The West Indians, who, by and large, came from a working-class background — they were mostly skilled craftsmen at this time — found it particularly difficult to accept their debarment from pubs and clubs and dance-halls (or to put up with the plangent racialism of the churches and/or their congregations). Fights broke out — and inevitably the police took the side of the whites. Gradually the West Indians began to set up their own clubs and churches and welfare associations — or met in barbers' shops and cafés and on street corners, as they were wont to do back home. The Indians and Pakistanis, on the other hand, were mostly rural folk and found their social life more readily in their temples and mosques and cultural associations. Besides, it was through these and the help of elders that the non-English speaking Asian workers could find jobs and accommodation, get their official forms filled in, locate their kinsmen or find their way around town.

In the area of work, too, resistance to racialism took the form of ad hoc responses to specific situations grounded in tradition. Often those responses were individualistic and uncoordinated, especially as between the communities — since Asians were generally employed in factories, foundries and textile mills, while recruitment of Afro-Caribbeans was concentrated in the service industries (transport, health and hotels). And even among these, there were 'ethnic jobs', like in the Bradford textile industry, and, often, 'ethnic shifts'.

A racial division of labour (continued more from Britain's colonial past than inaugurated in post-war Britain) kept the Asian and Afro-Caribbean workers apart and provided little ground for common struggle. Besides, the black workforce at this time, though concentrated in certain labour processes and areas of work, was not in absolute terms a large one — with West Indians outnumbering Indians and Pakistanis. Hence, the resistance to racial abuse and discrimination on the shop floor was more spontaneous than organised — but both individual and collective. Some workers left their jobs and went and found other work. Others just downed tools and walked away. On one occasion a Jamaican driver, incensed by the racialism around him, just left his bus in the High Street and walked off. (It was a tradition that reached back to his slave ancestry and would reach forward to his children.) But there were also efforts at collective action on the factory floor. Often these took the form of petitions and appeals regarding working conditions, facilities, even wages — but, unsupported by their white fellows, they had little effect. On occasions, there were attempts to form associations, if not unions, on the shop floor. In 1951, for instance, skilled West Indians in an ordnance factory in Merseyside (Liverpool) met secretly in the lavatories and wash rooms to form a West Indian Association which would take up cases of discrimination. But the employers soon found out and they were driven to hold their meetings in a neighbouring barber's shop — from which point the association became more community oriented. Similarly, in 1953 Indian workers in Coventry formed an association and named it, in the memory of Udham Singh, the Indian Workers' Association. But these early organisations generally ended up as social and welfare associations. The Merseyside West Indian Association, for instance, went through a period of vigorous political activity — taking up cases of discrimination and the cause of colonial freedom — but even as it grew in numbers and out of its barber's shop premises and into the white-run Stanley House, it faded into inter-racial social activity and oblivion.[2]

Discrimination in housing met with a community response from the outset: it was not, after all, a problem that was susceptible to individual solutions. Denied decent housing (or sometimes any housing at all), both Asians and Afro-Caribbeans took to pooling their savings till they were sizeable enough to purchase property. The Asians operated through an extended family system or 'mortgage clubs' and bought short-lease properties which they would rent to their kinsfolk and countrymen. Similarly, the West Indians operated a 'pardner' (Jamaican) or 'sou-sou' (Trinidadian) system, whereby a group of people (invariably from the same parish or island) would pool their savings and lend out a lump sum to each individual in turn. Thus their savings circulated among their own communities and did not go into banks or building societies to be lent out to white folk. It was a sort of primitive banking system engendered by tradition and enforced by racial discrimination. Of course, the prices the immigrants had to pay for the houses and the interest rates charged by the sources that were prepared to lend to them forced them into overcrowding and multi-occupation, invoking not only further racial stereotyping but, in later years, the rigours of the Public Health Act.

Thus it was around housing principally, but through traditional cultural and welfare associations and groups, that black self-organisation and self-reliance grew, unifying the respective communities. It was a strength that was to stand them in good stead in the struggles to come.

There was another area, too, where such organisation was significant — and offered up a different unity: the area of anti-colonial struggle. There had always been overseas students' associations — African, Asian, Caribbean — but in the period before the First World War these were mostly in the nature of friendship councils, social clubs or debating unions. But after that war and with the 'race riots' of 1919 (in Liverpool, London, Cardiff, Hull and other port areas where West African and *lascar* seamen had earlier settled) still fresh in their mind, West African students formed the West African Students' Union in 1925, with the explicit aim of opposing race prejudice and colonialism. It was followed in 1931 by the predominantly West Indian, League of Coloured Peoples. This was headed by an ardent Christian, Dr Harold Moody, and devoted to 'the welfare of coloured peoples in all parts of the world' and to 'the improvement of relations between the races'. But its journal, *Keys*, investigated and exposed cases of racial discrimination and in 1935, when 'colonial seamen and their families', especially in Cardiff, were being subjected to great economic hardship because of their colour, indicted 'the Trade Union, the Police and the Shipowners' of 'cooperating smoothly in barring coloured Colonial Seamen from signing on ships in Cardiff'.[3]

The connections — between colonialism and racialism, between black students and black workers — were becoming clearer, the campaigns more coordinated. And to this was added militancy when in 1937 a group of black writers and activists — including C.L.R. James, Wallace-Johnson, George Padmore, Jomo Kenyatta and Ras Makonnen — got together to form the International African Service Bureau. In 1944 the Bureau merged into the Pan-African Federation to become the British section of the Pan-African Congress Movement. From the outset, the Bureau (and then the Federation) was uncompromising in its demand for 'democratic rights, civil liberties and self-determination' for all subject peoples.

As the Second World War drew to a close and India's fight for *swaraj* stepped up, the movement for colonial freedom gathered momentum. Early in 1945 Asians, Africans and West Indians then living in Britain came together in a Subject Peoples'

Conference. Already in February that year, the Pan-African Federation, taking advantage of the presence of colonial delegates at the World Trade Union Conference, had invited them to a meeting at which the idea of another Pan-African Congress was mooted. Accordingly, in October 1945 the Fifth Pan-African Congress met in Manchester and, inspired by the Indian struggle for independence, forswore all 'gradualist aspirations' and pledged itself to 'the liquidation of colonialism and imperialism'. Nkrumah, Kenyatta, Padmore, James — they were names that were to crop up again (and again) in the history of anti-racist and anti-imperialist struggle.

Three years later India was free and the colonies of Africa and the Caribbean in ferment. By now, there was hardly an Afro-Caribbean association in Britain which did not espouse the cause of colonial independence and of black struggle generally. Asian immigrants, however, were past independence (so to speak) and the various Indian Leagues and Workers' Associations which had earlier taken up the cause of *swaraj* had wound down. In their place rose Indian Workers' Associations (the name was a commemoration of the past) concerned with immigrant issues and problems in Britain, though still identifying with political parties back home, the Communist Party and Congress in particular. So that two broad strands begin to emerge in IWA politics: one stressing social and welfare work and the other trade union and political activity — though not exclusively so.

In sum, the anti-racialist and anti-colonial struggle of this period was beginning to break down island and ethnic affiliations and associations and to re-form them in terms of the immediate realities of social and racial relations, engendering in the process strong community bases for the shop floor battles to come. But different interests predicated different unities and a different racialism engendered different though similar organisational impulses. There was no one unity — or two or three — but a mosaic of unities. However, as the colonies began to be free and the immigrants to become settled and the state to sanction and institute racial discrimination, and thereby provide the breeding ground for fascism, the mosaic of unities and organisations would resolve itself into a more holistic, albeit shifting, pattern of black unity and black struggle.

By 1955 the first 'wave' of immigration had begun to taper off: a mild recession had set in and the demand for labour had begun to drop (though London Transport was still recruiting skilled labour from Barbados in 1956). Left to itself, immigration from the West Indies was merely following the demand for labour; immigration from the Indian subcontinent, especially after the restrictions placed on it by the Indian and Pakistani governments in 1955, was now more sluggish. But racialism was hotting up and there were calls for immigration control, not least in the House of Commons. There had always been the occasional 'call', more for political reasons than economic; now the economy provided the excuse for politics. Pressure was also building up on the right. The loss of India and the impending loss of the Caribbean and Africa had spelt the end of empire and the decline of Britain as a great power. All that was left of the colonial enterprise was the ideology of racial superiority; it was something to fall back on. Mosley's pre-war British Union of Fascists was now revived as the Union Movement and was matched for race hatred by a rash of other organisations: A.K. Chesterton's League of Empire Loyalists, Colin Jordan's White Defence League, John Bean's National Labour Party, Andrew Fountaine's British National Party. And in the twilight area, between these and the right wing of the Tory Party, various societies for 'racial preservation' were beginning to sprout. Racial attacks became a regular part of immigrant life in Britain. More serious clashes occurred intermittently in London and several provincial cities. And in 1954, 'in a small street of terraced houses in Camden Town [London], racial warfare was waged for two days', culminating in a petrol bomb attack on the house of a West Indian.[4] Finally, in August 1958 large-scale riots broke out in Nottingham and were soon followed in Notting Hill (London), where teddy-boys, directed by the Mosleyites and the White Defence League, had for weeks gone on a jamboree of 'nigger-hunting' under the watchful eye of the police.

The blacks struck back, and even moderate organisations like the Committee of African Organisations, having failed to obtain 'adequate unbiased police protection', pledged to organise their own defence. The courts, in the person of a Jewish judge, Lord

Justice Salmon, made amends by sending down nine teddy-boys and establishing the right of 'everyone, irrespective of the colour of their skin... to walk through our streets with their heads erect and free from fear'. He also noted that the teddy-boys' actions had 'filled the whole nation with horror, indignation and disgust'. It was to prove the last time when such a claim could be made on behalf of the nation. Less than a year later a West Indian carpenter, Kelso Cochrane, was stabbed to death on the streets of Notting Hill. The police failed to find the killer. It was to prove the first of many such failures.

The stage was set for immigration control. But the economy was beginning to recover and the Treasury was known to be anxious about the prospect of losing a beneficial supply of extra labour for an economy in a state of expansion — though the ongoing negotiations for entry into the EEC promised another supply. Besides, the West Indian colonies were about to gain independence and moves towards immigration control, it was felt, should be postponed till after the British plan for a West Indian Federation had been safely established. Attempts to interest West Indian governments in a bilateral agreement to control immigration failed. In 1960 India withdrew its restrictions on emigration. In 1961 Jamaica withdrew its consent to a Federation. In early 1962 the Commonwealth Immigrants Bill was presented to parliament.

If the racial violence of Nottingham and Notting Hill had impressed on the West Indian community the need for greater organisation and militancy, the moves to impose 'coloured' immigration control strengthened the liaison between Asian and West Indian organisations. Already, in 1957, Claudia Jones, a Trinidadian and a communist, who had been imprisoned for her political activities in the US and then deported, had canvassed the idea of a campaigning paper. In March 1958 she, together with other West Indian progressives — Amy Garvey (the widow of Marcus Garvey) among them — brought out the first issue of what was to prove the parent Afro-Caribbean journal in Britain, the *West Indian Gazette*.[5] In 1959, after the Kelso Cochrane killing, Claudia Jones and Frances Ezzrecco (who had founded, in the teeth of the riots, the Coloured Peoples Progressive Association) led a deputation of West Indian organisations to the Home

Secretary. And in the same year, 'to get the taste of Notting Hill out of their throats', the *West Indian Gazette* launched the first Caribbean carnival in St Pancras Town Hall.

At about the same time, at the instance of the High Commission of the embryonic West Indian Federation (Norman Manley, Chief Minister of Jamaica, had flown to London after the troubles), the more moderate Standing Conference of West Indian Organisations in UK was set up. Although its stress was on integration and multi-racialism, it helped to cohere the island groups into a West Indian entity.

Nor had the Asian community been unmoved by the 1958 riots, for soon afterwards an Indian Association was formed in Nottingham and, more significantly, all the local IWAs got together to establish a central IWA-GB. (Nehru had advised it on his visit to London a year earlier.)

Now, with immigration control in the offing, other organisations began to develop — among them the Pakistani Workers' Association (1961) and the West Indian Workers' Association (1961). And these, along with a number of other Asian and Afro-Caribbean organisations, combined with sympathetic white groups to campaign against discriminatory legislation. The two most important umbrella organisations were the Co-ordinating Committee Against Racial Discrimination (CCARD) in Birmingham and the Conference of Afro-Asian-Caribbean Organisations (CAACO) in London. The former was set up in February 1962 by Jagmohan Joshi, of the IWA Birmingham, and Maurice Ludmer, an anti-fascist crusader from way back and later the founder-editor of *Searchlight*. CCARD itself had been inspired by a meeting at Digbeth called by the West Indian Workers' Association and the Indian Youth League to protest Patrice Lumumba's murder. That meeting had led to a march through Birmingham and other meetings against imperialism. In September 1961 CCARD led a contingent of blacks and whites through the streets of Birmingham in a demonstration against the Immigration Bill.

CAACO, initiated by the *West Indian Gazette* and working closely with the IWA and Fenner Brockway's Movement for Colonian Freedom, had its meetings and marches too, but it concentrated more on lobbying the High Commissions and parliament, particularly the Labour Party which had pledged to repeal the Act (if returned to power). But in August 1963, after the Bill had become Act and the Labour Party, with an eye to the elections, had begun to sidle out of its commitment, CAACO (with Claudia Jones at its head) organised a solidarity march from Notting Hill to the US embassy in support of 'negro rights' in the US and against racial discrimination in Britain — three days after Martin Luther King's People's March on Washington.

But international events also had adverse effects on black domestic politics. The Indo-China war in 1962 had split the communist parties in India. It now engendered schisms in the IWA-GB.

In April 1962 the Bill was passed and the battle lost. Racialism was no longer a matter of free enterprise; it was nationalised. If labour from the 'coloured' Commonwealth and colonies was still needed, its intake and deployment was going to be regulated not by the market forces of discrimination but by the regulatory instruments of the state itself. The state was going to say at the very port of entry (or non-entry) which blacks could come and which blacks couldn't — and where they could go and where they could live — and how they should behave and deport themselves. Or else... There was the immigration officer at the gate and the fascist within: racism was respectable, sanctioned, but with reason, of course; it was not the colour, it was the numbers — and for the immigrants' sakes — for fewer blacks would make for better race relations — and that, surely, must improve the immigrants' lot. It was a theme that was shortly to be honed to a fine respectability by Hattersley[6] and the Labour government. Evidently, hypocrisy too had to be nationalised. And in pursuit of that earnest, the Labour government of 1964 would make gestures towards anti-discriminatory legislation.[7]

Meanwhile, the genteel English 'let it all hang out'. In April 1963 the Bristol Omnibus Company discovered that it did not 'employ a mixed labour force as bus crews' — and freed from shame by the new absolution, it announced fearlessly 'a company may gain say fifteen coloured people and lose, through prejudice, thirty white people who decide they would sooner not work with them'.[8] But if Bristol — with three generations of black settlers and built on slavery — was only weighing up the statistics of prejudice, Walsall (with its more recent experience of blacks) made the more scientific pronouncement that 'coloureds can't react fast in traffic'. Bolton simply refused to engage 'riff-raff' any longer.[9]

The police felt liberated too. They had in the past appeared to derive only a vicarious pleasure from attacks on blacks; they had to be seen to be neutral. Now they themselves could go 'nigger-hunting' — the phrase was theirs — while officially polishing up on their neutrality. In December 1963 the British West Indian Association complained of increasing 'police brutality' stemming from the passing of the Commonwealth Immigrants Act. In 1964 the Pakistani community alleged that the wrists of Pakistani immigrants were being stamped with indelible ink at a police station in the course of a murder investigation: it was irrelevant that they had names and, besides, they all looked alike.[10] In 1965, WISC (the West Indian Standing Conference, which replaced the more moderate Standing Conference of West Indian Organisations in UK after the fall of the West Indian Federation in 1962) documented police excesses in Brixton and surrounding areas in a report on *Nigger-hunting in England*.[11] And at the ports of entry immigration officers, given carte-blanche in the 'instructions' handed down by the government, were having a field day.

At the local level, tenants' and residents' associations were organising to keep blacks out of housing. The number of immigrants had increased considerably in the two years preceding the ban: they were anxious to bring in their families and dependants before the doors finally closed. Housing, which had always been a problem since the war, became a more fiercely contested terrain. The immigrants had, of course, been consigned to slum houses and forced into multi-occupation. Now there were fears that they would move further afield into the white residential areas. At the same time, public health laws were invoked to dispel multi-occupation.

Schooling, too, presented a problem, as more and more 'coloured' children began to enter the country and sought places in their local schools. In October 1963 white parents in Southall (which had a high proportion of Indians) demanded separate classes for their children because 'coloured' children were holding back their progress. In December the Commonwealth Immigrants' Advisory Council (CIAC), which had been set up to advise the Home Secretary on matters relating to 'the welfare and integration of immigrants', reported that 'the presence of a high proportion of immigrant children in one class slows down the general routine of working and hampers the progress of the whole class, especially where the immigrants do not speak or write English fluently'. This, it said, was bad for immigrant children too — for 'they would not get as good an introduction to British life as they would get in a normal school'. Besides, there was the danger of white parents removing their children and making some schools 'predominantly immigrant schools'.[12] In November Sir Edward Boyle, the avowedly liberal Minister of Education, told the House of Commons that it was 'desirable on education grounds that no one school should have more than about 30 per cent of immigrants'. Accordingly, in June 1965 Boyle's law enacted that there should be no more than a third of immigrant children in any school; the surplus should be bussed out[13] — but white children would not be bussed in.

As for West Indian children, whose difficulties were ostensibly 'Creolese English', low educability and 'behaviour problems', the solution would be found in 'remedial classes' and even 'special' schools.

None of these measures — or the various instances of discrimination — went without protest, however. The Bristol Bus Company, for instance, was subjected to a boycott and demonstrations for weeks until it finally capitulated. Police harassment, as already mentioned, was documented and publicised by West Indian organisations. The relegation of West Indian children into special classes and/or schools was fought — and continued to be fought — first by the North London West Indian Association and then by other local and national organisations.

But, by and large, the unity — between West Indians and Asians, militants and moderates — that had sprung up between the riots of 1958 and the Immigration Act of 1962 had been dissipated by more immediate concerns. Now there were families to house and children to school and dependants to look after:

the immigrants were becoming settlers. And since it was Asian immigrants who, more than the West Indian, had come on a temporary basis — to make enough money to send to their impoverished homes — before the Immigration Act foreclosed on them, it was their families and relatives who swelled the numbers now. And their politics tended to become settler politics — petitioning, lobbying, influencing political parties, weighing-in on (if not yet entering) local government elections — and their struggles working-class struggles on the factory floor — and they, by virtue of being fought in the teeth of trade union racism, were to prove political too.

In May 1965 the first important 'immigrant' strike took place — at Courtauld's Red Scar Mill in Preston — over the management's decision to force Asian workers (who were concentrated, with a few West Indians, in one area of the labour process) to work more machines for (proportionately) less pay.[14] The strike failed, but not before it had exposed the active collaboration of the white workers and the union with management. A few months earlier, a smaller strike of Asian workers at Rockware Glass in Southall (London) had exposed a similar complicity. And the Woolf Rubber Company strike later in the same year, though fought valiantly by the workers, supported by the Asian community and in particular by the IWA, lost out to the employers through lack of official union backing.[15]

The Afro-Caribbean struggles of this period (post 1962) also reflected a similar community base, though different in origins. Ghana had become free in 1957, Uganda in 1962 and Jamaica and Trinidad and Tobago in the same year, but there were other black colonies in Africa and the Caribbean still to be liberated. And then there was the black colony in North America, which, beginning with Martin Luther King's civil rights movement, was revving up into the Black Power struggles of the mid-1960s.

King visited London on the way to receiving his Nobel prize in Oslo in December 1964. And at his instigation, a British civil rights organisation, the Campaign Against Racial Discrimination (CARD) was formed in February 1965 — federating various Asian and Afro-Caribbean organisations and sporting Labour Party 'radicals'.

More significant, however, was the visit of Malcolm X. Malcolm blitzed London in February 1965 and in his wake was formed a much more militant organisation, the Racial Action Adjustment Society (RAAS)[16] with Michael de Freitas, later Abdul Malik and later still Michael X, at its head. 'Black men, unite,' it called, 'we have nothing to lose but our fears.'

It is the fashion today, even among blacks, to see Michael X only as a criminal who deserved to be hanged for murder by the Trinidadian government (1975). (The line between politics and crime, after all, is a thin one — in a capitalist society.) But it was Michael and Roy Sawh (a Guyanese 'Indian') and their colleagues in RAAS who, as we shall see, more than anybody else in this period freed ordinary black people from fear and taught them to stand up for their rights and their dignity.[17] It was out of RAAS, too, that a number of our present-day militants have emerged.

It is also alleged, in hindsight and contempt, that RAAS had no politics but the politics of thuggery. But it was RAAS who descended on Red Scar Mills in Preston to help the Asian strikers (at their invitation). And the point is telling — if only because it marked a progression in the *organic* unity of the (Afro-) Asian, 'coolie', and (Afro-) Caribbean, slave, struggles in the diaspora, begun in Britain by Claudia Jones' *West Indian Gazette and Afro-Asia-Caribbean News* on which Abimanyu Manchanda, an Indian political activist and a key figure in the British anti-Vietnam war movement, was to work.

But RAAS, or black militancy generally, would not have had the backing it did but for the growing disillusion with the Labour Party's policies on immigration control and, therefore, racism. The 'coloured immigrants' still had hopes in the party of the working class and of colonial independence and sought to influence its policies. But after the 1964 general election, Labour's position became clearer. Peter Griffiths, the Tory candidate for Smethwick (an 'immigrant area' in Birmingham), had campaigned on the basis of ending immigration and repatriating 'the coloureds'.

> If you want a nigger neighbour
> Vote Labour

he had sloganised — and won. But Labour won the election and

Harold Wilson, the incoming Prime Minister, denounced Griffiths as 'a parliamentary leper'. However, Wilson's policies were soon to become leprous too: the Immigration Act was not only renewed in the White Paper of August 1965 but went on to restrict 'coloured immigration' further on the basis that fewer numbers made for better race relations. In pursuit of that philosophy, Labour then proceeded to pass a Race Relations Act (September 1965), which threatened racial discrimination in 'places of public resort' with conciliation. It was prepared, however, to penalise 'incitement to racial hatred' — and promptly proceeded to prosecute Michael X.

Equally off-target and ineffectual were the two statutory bodies that Labour set up, the National Committee for Commonwealth Immigrants (NCCI) and the Race Relations Board (RRB) — the one chiefly to liaise with immigrants and ease them out of the difficulties (linguistic, educational, cultural and so on) which prevented integration, and the other, as mentioned above, to conciliate discrimination in hotels and places through conciliation committees.

To ordinary blacks these structures were irrelevant: liaison and conciliation seemed to define them as a people apart who somehow needed to be fitted into the mainstream of British society — when all they were seeking were the same rights as other citizens. They (liaison and conciliation) were themes that were to rise again in the area of police-black relations — this time as substitutes for police accountability — and not without the same significance.

But if NCCI failed to integrate the 'immigrants', it succeeded in disintegrating 'immigrant' organisations — the moderate ones anyway and local ones mostly — by entering their areas of work, enticing local leaders to cooperate with them (and therefore government) and pre-empting their constituencies. Its greatest achievement was to lure the leading lights of CARD into working with it, thereby deepening the contradictions in CARD (as between the militants and the moderates). WISC and NFPA[18] disaffiliated from CARD and the more militant blacks followed suit, leaving CARD to its more liberal designs. The government had effectively shut out one area of representative black opinion. But an obstacle in the way of the next Immigration Act had been cleared.

When CARD finally broke up in 1967 the press and the media generally welcomed its debacle. They saw its sometimes uncompromising stand against racial discrimination as a threat to 'integration' (if not to white society), resented its outspoken and articulate black spokes-men and women, denouncing them as communists and maoists, and feared that it would emerge from a civil rights organisation into a Black Power movement. The 'paper of the top people', having warned the nation of 'The Dark Million' in a series of articles in the months preceding the August 1965 White Paper, now wrote, 'there are always heavy dangers in riding tigers — and these dangers are not reduced when the animal changes to a black panther' (*The Times*, 9 November 1967). And a *Times* news team was later to write that 'the ominous lesson of CARD... is that the mixture of pro-Chinese communism and American-style Black Power on the immigrant scene can be devastating'.[19] (International events were beginning to cast their shadows.)

The race relations pundits added their bit. The Institute of Race Relations (some of whose Council members and staff were implicated in CARD politics) commissioned a book giving the liberal version. Although an independent research organisation, replete with academics, the IRR had already been moving closer to government policies on immigration and integration — backing them with 'objective' findings and research.[20]

The race scene was changing — radically. The Immigration Acts, whatever their racialist promptings, had stemmed from an economic rationale, fashioned in the matrix of colonial-capitalist practices and beliefs. They served, as we have seen, to take racial discrimination out of the market-place and institutionalise it — inhere it in the structures of the state, locally and nationally. So that at both local and national levels 'race' became an area of contestation for power. It was the basis on which local issues of schooling and housing and jobs were being, if not fought, side-tracked. It was an issue on which elections were won and lost. It was an issue which betrayed the trade unions' claims to represent the whole of the working class, and so betrayed the class. It had entered the arena of politics (not that, subliminally, it was not always there) and swelled into an ideology of racism to be borrowed by the courts in their decision-making and by the fascists for their regeneration.[21]

Racial attacks had already begun to mount. In 1965, in the months preceding the White Paper but after Griffiths' victory, 'a Jamaican was shot and killed... in Islington, a West Indian schoolboy in Notting Hill was nearly killed by white teenagers armed with iron bars, axes and bottles... a group of black men outside a café in Notting Hill received blasts from a shot-gun fired from a moving car', hate leaflets appeared in Newcastle-upon-Tyne, crosses were burnt outside 'coloured citizens' homes' in Leamington Spa, Rugby, Coventry, Ilford, Plaistow and Cricklewood and 'a written warning (allegedly) from the Deputy Wizard of the Ku Klux Klan was sent to the Indian secretary of CARD: "You will be burnt alive if you do not leave England by August 31st".'[22] The British fascists, however, denied any connection with the Klan — not, it would appear, on a basis of fact, but in the conviction that there was now a sufficient ground-swell of grassroots racism to float an electoral party. Electoral politics, of course, were not going to bring them parliamentary power, but they would provide a vehicle for propaganda and a venue for recruitment — and all within the law. They could push their vile cause to the limits of the law, within the framework of the law, forcing the law itself to become more repressive of democratic freedoms. By invoking their democratic right to freedom of speech, of association, etc. — by claiming equal TV time as the other electoral parties and by gaining 'legitimate' access to the press and radio — they would propagate the cause of denying others those freedoms and legitimacies, the blacks in the first place. They would move the whole debate on race to the right and force incoming governments to further racist legislation — on pain of electoral defeat. And so, in February 1967 the League of Empire Loyalists, the British National Party and local groups of the Racial Preservation Society merged to form the National Front (NF) — and in April that year put up candidates for the Greater London Council elections.

But they — and the government — reckoned without the blacks. The time was long gone when black people, with an eye to returning home, would put up with repression: they were settlers now. And state racism had pushed them into higher and more militant forms of resistance — incorporating the resistances of the previous period and embracing both shop floor and community, Asians and Afro-Caribbeans, sometimes in different areas of struggle, sometimes together.

RAAS, as we have seen, was formed in 1965. It was wholly an indigenous movement arising out of the opposition to native British racism but catalysed by Malcolm X and the Black Muslims. And so it took in, on both counts, the African, Asian, Afro-Caribbean and Afro-American dimensions of struggle and the struggles in the workplace and the community. It had, almost as its first act, descended on the Red Scar Mills in Preston to help the Asian strikers. It then set up office in a barber's shop in Reading and worked and recruited in the North and the Midlands with Abdulla Patel (one of the strikers from Red Scar Mills) and Roy Sawh as its organisers there. In London, too, RAAS gathered a sizeable following through its work with London busmen and its legal service (Defence) for black people in trouble with the police. It was written up in the press, often as a novel and passing phenomenon, and appeared (in a bad light) on BBC's Panorama programme. The disillusion with CARD swelled RAAS's numbers. And the indictment of Michael X in 1967 for 'an inflammatory speech against white people' — when white people indulged freely in racist abuse — served to validate RAAS's rhetoric. At Speakers' Corner Roy Sawh and other black speakers would inveigh against 'the white devil' and 'the Anglo-Saxon swine', and find a ready and appreciative audience.

In June 1967 the Universal Coloured Peoples' Association (UCPA) was formed, headed by a Nigerian playwright, Obi Egbuna. It too arose from British conditions, but, continuing in the tradition of the struggle against British colonialism, stressed the need to fight both imperialism and racism. The anti-white struggle was also anti-capitalist and anti-imperialist — universal to all coloured peoples. And so its concerns extended from racism in Britain and elsewhere to the war in Vietnam, the independence of Zimbabwe, the liberation of 'Portuguese Africa', the cultural revolution in China. It was, of course, inspired and influenced by the Black Power struggle in America and, more immediately, by Stokely Carmichael's visit to London in July that year.

'Black Power', Egbuna declared at a Vietnam protest rally in Trafalgar Square in October 1967, 'means simply that the blacks of this world are out to liquidate capitalist oppression of black people wherever it exists by any means necessary.' Black people in Britain, the UCPA pointed out, though numerically small, were so concentrated in vital areas of industry, hospital services (a majority of doctors and nurses in the conurbations were black) and transport that a black strike could paralyse the economy. Some UCPA speakers at meetings in Hyde Park urged more direct action. Roy Sawh (the members of one organisation were often members of another) 'urged coloured nurses to give wrong injections to patients, coloured bus crews not to take the fares of black people... [and] Indian restaurant owners to "put something in the curry" '.[23] Alex Watson, a Jamaican machine-operator, was reported to have exhorted coloured people to destroy the whites.[24] Ajoy Ghose, an unemployed Indian, pointed out that to kill whites was not murder and Uyornumu Ezekiel, Nigerian electrician, having derided the prime minister as a 'political prostitute', said that England was 'going down the toilet'.[25] They were all prosecuted; Ezekiel was discharged, the others fined.

But UCPA rhetoric was helping to stiffen black backs, its meetings and study groups to raise black consciousness, its ideology to politicise black people. The prosecution of its members showed up the complicity of the courts — 'protection rackets for the police', the secretary of WISC was to call them. And its example, like that of RAAS, encouraged other black organisations to greater militancy.

RAAS, it would appear, stressed black nationalism, while the UCPA emphasised the struggles of the international working class. But they were in fact different approaches to the same goals. RAAS's 'nationalism', stemming as it did from the West Indian experience, combined an understanding of how colonialism had divided the Asian and African and Caribbean peoples (coolie, savage and slave) with an awareness of how that same colonialism made them one people now: they were all blacks. Hence the Black House (and the cultural groups) that RAAS was briefly to set up on 1970 did not, like Leroi Jones/Amiri Baraka's Spirit House which was its inspiration, exclude other 'coloureds'; the historical experience was different.

Meanwhile, the number of black strikes — mainly Asian, because it was they who were employed in the menial jobs in the foundries, the textile and paper mills, the rubber and plastic works — began to mount, and nearly all of them showed up trade union racism. Working conditions in the foundries were particularly unendurable. The job itself involved working with molten metal at 1400 degrees centigrade. Burns and injuries were frequent. The wage was rarely more than £14 a week and promotion to skilled white-only jobs was unthinkable. Every action on the part of the Asian workers was either unsupported or opposed by the trade union to which they belonged, for example in the strike at Coneygre Foundry (Tipton) in April 1967 and again in October 1968, at the Midland Motor Cylinder Co in the same year and at Newby Foundry (West Bromwich) a year later. But the support from their communities and community organisations was unwavering. The temples gave free food to strikers, the grocers limitless credit, the landlords waived the rent. And joining in the strike action were local organisations and associations — IWAs and Pakistani Welfare Associations and/or other black organisations or individuals connected with them.

Some issues, however, embraced the whole community more directly. For, apart from the general question of wages, conditions of work, etc., quite a few of these strikes also involved 'cultural' questions, such as the right to take time off for religious festivals, the right to break off for daily prayer (among Moslems), the right of Sikh busmen to wear turbans instead of the official head-gear. And because of trade union opposition to such 'practices', the struggles of the class and the struggles of the community, of race, became indistinguishable.

These in turn were linked to the struggles back 'home' in the subcontinent — if only through family obligations arising from economic need predicated by under-development caused by imperialism. The connections were immediate, palpable, personal. Imperialism was not a thing apart, a theoretical concept; it was a lived experience — only one remove from the experience of racism itself. And for that reason, too, the politics and political organisations of the 'home' countries had a bearing on the life and politics of Indian and Pakistani settlers in Britain — now not in terms of electoral political parties so much as in terms of the resistance movements to the authoritarian state[26] — which, in turn, had resonances for them in Britain. Asian-language newspapers kept them in constant touch with events in the subcontinent, and the political refugees whom they housed and looked after not only involved them in their movements, but fired their own resistances. Reciprocally, their people back 'home' were keened to the mounting racism in Britain.

On all these fronts, then, blacks by 1968 were beginning to fight as a class and as a people. Whatever the specifics of resistance in the respective communities and however different the strategies and lines of struggle, the experience of a common racism and a common fight against the state united them at the barricades. The mosaic of unities observed earlier resolved itself, before the onslaught of the state, into a black unity and a black struggle. It would recede again when the state strategically retreated into urban aid programmes and the creation of a class of black collaborators — only to be forged anew by another generation, British-born but not British.

In March 1968 the Labour government passed the 'Kenyan Asian Act', this time barring free entry to Britain of its citizens in Kenya — because they were Asians. O.K. so they held British passports issued by or on behalf of the British government, but that did not make them British, did it? Now, if they had a parent or grandparent 'born, naturalised, or adopted in the UK' — like those chaps in Australia, New Zealand and places — it would be a different matter. But of course the government would set aside a quota of entry vouchers especially for them — for the British Asians, that is. The reasons were not 'racial', as Prime Minister Wilson pointed out to the Archbishop of Canterbury, but 'geographical'.[27]

Between conception and passage the Act had taken but a week. The orchestration of public opinion that preceded it had gone on for about a year, but it had risen to a crescendo in the last six months. In October Enoch Powell, man of the people, had warned the nation that there were 'hundreds of thousands of people in Kenya'[28] who thought they belonged to Britain 'just like you and me'. In January the press came out with scare stories, as it had done before the Immigration Act of 1962, except this time it was not smallpox but the clandestine arrival of hordes of Pakistanis. In February Powell returned to his theme — and other politicians joined in. Later in the month the *Daily Mirror*, the avowedly pro-Labour paper, warned of an 'uncontrolled flood of Asian immigrants from Kenya'. On 1 March the bill was passed.

Blacks were enraged. They had lobbied, petitioned, reasoned, demonstrated — even campaigned alongside whites in NCCI's Equal Rights set-up — and had made no impact. But the momentum was not to be lost. Within weeks of the Act Jagmohan Joshi, Secretary of the IWA Birmingham, was urging black organisations to form a broad, united front. On 4 April Martin Luther king was murdered...

'I have a dream.' They slew the dreamer.

Some two weeks later Enoch Powell spoke of his and the nation's nightmare: the blacks were swarming all over him, no, all over the country, 'whole areas and towns and parts of towns across England' were covered with them, they pushed excreta through old ladies' letter boxes; we must take 'action now', stop the 'inflow', promote the 'outflow', stop the fiancés, stop the dependants, 'the material of the future growth of the immigrant-descended population', the breeding ground. 'Numbers are of the essence.'

What Powell says today, the Tories say tomorrow and Labour legislates on the day after. Immediately, it was public opinion that was roused. The press picked up Powell's themes and Powell. The unspeakable had been spoken, free speech set free, the whites liberated; Asians and West Indians were abused and attacked, their property damaged, their women and children terrorised. Police harassment increased, the fascists went on a rampage and Paki-bashing emerged as the national sport. A few trade unionists made gestures of protest and earned the opprobrium of the rank and file. White workers all over the country downed tools and staged demonstrations on behalf of Powell. And on the day that even-handed Labour, having passed a genuinely racist Immigration Act, was debating a phoney anti-racist Race Relations Bill, London dockers struck work and marched on parliament to demand an end to immigration. Three days later they marched again, this time with the Smithfield meat-porters.

But the blacks were on the march too. On the same day as the dockers and porters marched, representatives from over fifty organisations (including the IWAs, WISC, NFPA, UCPA, RAAS, etc.) came together at Leamington Spa to form a national body, the Black People's Alliance (BPA), 'a militant front for Black Consciousness and against racialism'. And in that, the BPA was uncompromising from the outset. It excluded from membership immigrant organisations that had compromised with government policy or fallen prey to government hand-outs (Labour's Urban Aid programme was beginning to percolate through to the blacks) or looked to the Labour Party for redress. For, in respect of 'whipping up racial antagonisms and hatred to make political gains', there was no difference between the parties or between them and Enoch Powell. He was 'just one step in a continuous campaign' which had served to give 'the green light to the overtly fascist organisations... now very active in organising among the working class'.[29] Member organisations would continue to maintain their independent existence and function at the local level, in terms of the particular communities and problems; the BPA would operate on the national level, coordinating the various fights against state racism. And, where necessary, it would take to the streets en masse — as it did in January 1969 (during the Commonwealth Prime Ministers' Conference), when it led a march of over 7,000 people to Downing Street to demand the repeal of the Immigration Acts.

From Powell's speech and BPA, but nurtured in the Black Power movement, sprang a host of militant black organisations all over the country, with their own newspapers and journals, taking up local, national and international issues. Some Jamaican organisations marched on their High Commission in London, protesting the banning of the works of Stokely and Malcolm X in Jamaica, while others, like WISC, RAAS and the Caribbean Artists' movement, sent petitions. On the banning of Walter Rodney from returning to his post at the university, Jamaicans staged a sit-in at the High Commission. A 'Third

World' Benefit for three imprisoned playwrights — Wole Soyinka in Nigeria, LeRoi Jones in America and Obi Egbuna in Britain — was held at the Round House with Sammy Davis, Black Eagles and Michael X. But Egbuna's imprisonment (with two other UPCA members), for uttering and writing threats to kill police, also stirred up black anger. 'Unless something is done to ensure the protection of our people,' wrote the Black Panther Movement in its circular of 3 October 1968, 'we will have no alternative but to rise to their defence. And once we are driven to that position, redress will be too late, Detroit and Newark will inevitably become part of the British scene and the Thames foam with blood sooner than Enoch Powell envisaged.'[30]

Less than six months after Powell's speech, Heath, the Tory leader, having sacked Powell from the Shadow Cabinet, himself picked up Powell's themes. Immigration, he said, whether of voucher-holders or dependants, should be 'severely curtailed' — and those who wished to return to their country of origin should 'receive assisted passage from public funds'. But Powell outbid him in a call for a 'Ministry of Repatriation' and 'a programme of large-scale voluntary but organised, financed and subsidised repatriation and re-emigration'. Two months later Heath upped the ante: the government should stop all immigration. Powell, who was on the same platform, applauded him. Callaghan, the Home Secretary, however, derided Heath's speech as 'slick and shifty'. Three days later Callaghan debarred Commonwealth citizens from entering Britain to marry fiancées and settle here, 'unless there are compassionate circumstances'. And in May 1969, in an even more blatant piece of 'even-handedness', Callaghan sneaked into the 'liberal' Immigration Appeals Bill[31] a clause which stipulated that dependants should henceforth have entry certificates before coming to Britain. And with that he set the seal on the prevarications, delays and humbug that British officials in the countries of emigration, mainly India and Pakistan now, subjected dependants to — till the young grew old in the waiting and old folk just gave up or died. It was a move in Powell's direction, but he meanwhile had moved on to higher things, like calculating the cost of repatriation. Before Callaghan could move towards him again, Labour lost the election (1970). It was now left to a Tory government under Heath to effect Powellite policies (on behalf of Labour) in the Immigration Act of 1971.

The new Act stopped all primary (black) immigration dead. Only 'patrials' (Callaghan's euphemism for white Commonwealth citizens) had right of abode now. Non-patrials could only come in on a permit to do a specific job in a specific place for a specific period. Their residence, deportation, repatriation and acquisition of citizenship were subject to Home Office discretion. But constables and immigration officers were empowered to arrest without warrant anyone who had entered or was suspected ('with reasonable cause') to have entered the country illegally or overstayed his or her time or failed to observe the rules of the Act in any other particular. Since all blacks were, on the face of them, non-patrials, this meant that all blacks were illegal immigrants unless proved otherwise. And since, in this respect, the Act (when it came into force in January 1973) would be retrospective, illegal immigrants went back a long way.

Entry of dependants of those already settled in Britain would continue to be made on the basis of entry certificates issued by the British authorities, at their discretion, in the country of emigration. Eligibility — as to age, dependancy, relationship to the relative in the UK, etc. — would have to be proved by the dependant. Children would have to be under 18 to be eligible at all and parents over 65. But 'entry clearances' did not guarantee entry into Britain. It could still be refused at the port of entry by the immigration officer on the ground that 'false representations were employed or material facts were concealed, *whether or not to the holder's knowledge*, for the purpose of obtaining clearance' [emphasis added]. (When, in 1980, Filipino domestic workers who had entered legally were to ask to bring in their children, they would be deported — on the basis of this clause — for having withheld information (re children) which they were not asked for in the first place.) And as for those who wanted to be repatriated, every assistance would be afforded.

On the face of it, the Act appeared no more racist than its predecessors. Bans and entry certificates, stop and search arrests and 'Sus' (under section 4 of the 1824 Vagrancy Act anyone could be arrested 'on suspicion' of loitering with intent to commit an arrestable offence), detention and deportations were already everyday aspects of black life. Even the distinction that the Act made between the old settlers and the new migrants to make them all migrants again did not seem to matter much: they had never been anything but 'coloured immigrants'. But there was something else in the air. The 'philosophy' had begun to change, the raison d'être of racism. It was not that racism did not make for cheap labour any more, but that there was no need for capital to import it. Instead, thanks to advances in technology and changes in its own nature, capital could now move to labour, and did — the transnational corporations saw to that.[32] The problem was to get rid of the labour, the black labour that was already here. And racism could help there — with laws and regulations that kept families apart, sanctioned police harassment, invited fascist violence and generally made life untenable for the black citizens of Britain. And if they wanted to return 'home', assisted passages would speed their way.

To get the full flavour of the Immigration Act of 1971, however, it must be seen in conjunction with the Industrial Relations Act of the same year. For if the Immigration Act affected the black peoples (in varying ways), the Industrial Relations Act, which put strictures on trade unions and subjected industrial disputes to the jurisdiction of a court, the National Industrial Relations Court (NIRC), affected the black working class specifically. As workers, they were subject to the Industrial Relations Act's overall attack on the class (and later to the government's three-day week). As blacks, they were subject to the Immigration Act's threat of deportation — as illegal immigrants or for acting in ways not 'conducive to the public good'. As blacks and workers, they were subjected to the increasing racism of white workers and trade unions under siege — and more susceptible to being offered up to the NIRC for the adjudication of their disputes. Together, the Acts threatened to lock the black working class into the position of a permanent under-class. Hence, it is precisely in the area of black working-class struggles that the resistance of the early 1970s becomes significant.

But these were not struggles apart. They were, because they were black, tied up with other struggles in the community, which in turn was involved in the battles on the factory floor. The community struggles themselves had, as we have seen, become increasingly politicised in the Black Power movement and organised in black political groups. And they, after the failure of the white left to acknowledge the special problems of the black working class or the need for black self-help and organisation, began to address themselves to the problems of black workers — in the factories, in the schools, in their relationship with the police. Which in turn was to lead to more intense confrontation with, if not the state directly, the instruments of state oppression. But since these operated differentially in respect of the Asian and West Indian communities, the resistance to them was conducted at different levels, in different venues, with (often) different priorities.

The energies of the Asian community, for instance, were taken up with trying to get their families and dependants in — and once in, to keep them (and themselves) from being picked up as illegal immigrants. Since these required a knowledge of the law and of officialdom, it was inevitable that their struggles in this respect would be channelled into legal battles — mainly through the Joint Council for the Welfare of Immigrants (JCWI)[33] with its expertise and commitment — and into petitioning and lobbying. This aspect was further reinforced by the issue of the 'shuttlecock Asians', those British Asians in East Africa who (for one reason or another) were bandied about from country to country before eventually being imprisoned in Britain prior to admission.[34] From 1972 Asian leaders and organisation were also preoccupied with the resettlement of British Asian refugees from Amin's Uganda.

If the struggles to gain entry for their families and dependants drained the energies of the Asian community at one level, the abuse and the humiliation that those seeking entry were submitted to by immigration officials served to degrade and sometimes demoralise it. The instances are legion and have been documented elsewhere.[35] But the most despicable of them all was the vaginal examination of women for virginity — in itself an appalling violation but, for women from a peasant culture, a violation beyond violence. Further debilitating of the community was the police use of informers to apprehend suspected illegal immigrants individually and through 'fishing raids', generating thereby suspicion and distrust among families. In turn, the battles against them channelled the community's energies into getting the retrospective aspect of the Immigration Act regarding illegal immigrants repealed (and was finally 'rewarded' by the dubious amnesty of 1974 for all those who had entered illegally before 1973). But the police's Illegal Immigration Intelligence Unit (IIIU) remained in force.[36]

The Afro-Caribbean community, for its part (excepting the workplace, which will be treated separately), was occupied with fighting the mis-education of its children and the harassment of the police. Both problems had existed before, but they had now gathered momentum. West Indian children were consistently and right through the schooling system treated as uneducable and as having 'unrealistic aspirations' together with a low IQ. Consequently, they were 'banded' into classes for backward children or dumped in ESN (educationally subnormal) schools and forgotten. The fight against categorisation of their children as under-achieving, and therefore fit only to be an under-class, begun in Haringey (London) in the 1960s by West Indian parents, teachers and the North London West Indian Association (NLWIA) under Jeff Crawford, now spread to other areas and became incorporated in the programmes of black political organisations. An appeal to the Race Relations Board (1970) elicited the response that the placement of West Indian children in ESN schools was 'no unlawful act'. The Caribbean Education Association then held a conference on the subject and in the following year Bernard Coard (now Deputy Prime Minister of Grenada) wrote his influential work, *How the West Indian child is made educationally subnormal...* Black militants and organisations, meanwhile, had begun to set up supplementary schools in the larger conurbations. In London alone there was the Kwame Nkrumah school (Hackney black teachers), the Malcolm X Montessori Programme (Ajoy Ghose), the George Padmore school (John La Rose[37] and the Black Parents' Movement), the South-east London Summer School (BUFP: Black Unity and Freedom Party), Headstart (BLF: Black Liberation Front) and the Marcus Garvey school (BLF and others).[38]

Projects were also set up to teach skills to youth. The Mkutano Project, for instance, started by the BUFP (in 1972), taught typing, photography, Swahili; the Melting Pot, begun about the same time by Ashton Gibson (once of RAAS), had a workshop for making clothes; and Keskidee, set up by an ex-CARD official, Oscar Abrams, taught art and sculpture and encouraged black poets and playwrights. For older students, Roy Sawh ran the Free University for Black Studies. And then there were hostels for unemployed and homeless black youth — such as Brother Herman's Harambee and Vince Hines' Dashiki (both of whom had been active in RAAS) — and clubs and youth centres. Finally, there were the bookshop cum advice centres, such as the Black People's Information Centre, BLF's Grassroots Storefront and BWM's Unity Bookshop,[39] and the weekly or monthly newspapers: *Black Voice* (BUFP), *Grassroots* (BLF), *Freedom News* (BP: Black Panthers), *Frontline* (BCC: Brixton and Croydon Collective), *Uhuru* (BPFM: Black People's Freedom Movement), *BPFM Weekly*, the *BWAC Weekly* (Black Workers' Action Committee) and the less frequent but more theoretical journal *Black Liberator* — and a host of others that were more ephemeral. Some of these papers took up the question of black women, and the BUFP, following on the UCPA's Black Women's Liberation Movement, had a black women's action committee.

RAAS's Black House was going to be a huge complex, encompassing several of these activities. But hardly had it got off the ground in February 1970 than it was raided by the police and closed down. And RAAS itself began to break up. Members of RAAS, however, went on to set up various self-help projects — as indicated above.

By 1971 the UCPA was also breaking up into its component groups, with the hard core of them going to form the BUFP. (National bodies were by now not as relevant to the day-to-day struggles as local ones and the former's unifying role could equally be fulfilled by ad hoc alliances.) The UCPA, RAAS, the Black Panthers and other black organisations had in the previous two years been increasingly occupied with the problem of police brutality and fascist violence. The success of Black Power had brought down on its head the wrath of the system. Its leaders were persecuted, its meetings disrupted, its places of work destroyed. But it had gone on gaining momentum and strength: it was not a party, but a movement, gathering to its concerns all the strands of capitalist oppression, gathering to its programme all the problems of oppressed peoples. There was hardly a black in the country that did not identify with it and, through it, to all the non-whites of the world, in one way or another. And as for the British-born youth, who had been schooled in white racism, the movement was the cradle of their consciousness. Vietnam, Guinea-Bissau, Zimbabwe, Azania were all their battle-lines, China and Cuba their exemplars. The establishment was scared. The media voiced its fears. There were rumours that Black Power was about to take over Manchester City Council.[40]

In the summer of 1969 the UCPA and the Caribbean Workers' Movement were documenting and fighting the cases of people beaten up and framed by the police — in Manchester and London. In August the UCPA held a Black Power rally against 'organised police brutality' on the streets of Brixton. In April 1970 the UCPA and the Pakistani Progressive Party staged a protest outside the House of Commons over 'Paki-bashing' in the East End of London. And the Pakistani Workers' Union called for citizens' defence patrols: a number of Asians had been murdered in 1969 and 1970. A month later over 2,000 Pakistanis, Indians and West Indians marched from Hyde Park to Downing Street demanding police protection from skinhead attacks. In the summer of 1970 police attacks on blacks — abuse, harassment, assaults, raids, arrests on 'Sus', etc., in London, Manchester, Bristol, Birmingham, Leeds, Liverpool, etc. — put whole black communities under siege. In July and August there were a series of clashes between black youth and the police in London and on one occasion over a hundred youth surrounded the Caledonian Road police station demanding the release of four blacks who had been wrongfully arrested. Things finally came to a head in Notting Hill on 9 August, when police broke up a demonstration against the proposed closure of the Mangrove restaurant with unprecedented violence. The blacks fought back, a number of them were arrested and nine of the 'ringleaders' subsequently charged with riot, affray and assault.

The Mangrove was a meeting place and an eating place, a social and welfare club, an advice and resource centre, a black house for black people, a resting place in Babylon. And if only for this reason, the police could not leave it alone. They raided it and raided it, harassed its customers and relentlessly persecuted its owner, Frank Critchlow. They made it the test of police power; the blacks made it a symbol of resistance. The battle of the blacks and the police would be fought over the Mangrove.

The trial of the Mangrove 9 (October-December 1971) is too well documented to be recounted here, but, briefly, they won. They did more: they took on what the defence counsel called 'naked judicial tyranny' — some by conducting their own defence — and won. Above all, they unfolded before the nation the corruption of the police force, the bias of the judicial system, the racism of the media — and the refusal of black people to submit themselves to the tyrannies of the state. Other trials would follow and even more bizarre prosecutions be brought, as when the alleged editor of *Grassroots* was charged with 'encouraging the murder of persons unknown' by reprinting an article from the freely available American Black Panther paper on how to make Molotov cocktails. But they would all be defended — by the whole community — and become another school of political education.[41]

If the Mangrove marked the high water-mark of Black Power and lowered the threshold of what black people would take, it also marked the beginnings of another resistance: of black youth condemned by racism to the margins of existence and then put upon by the police. 'Sus' had always laid them open to police harassment, but the government's White Paper on Police-Immigrant Relations in 1973, which warned of 'a small minority of young coloured people... anxious to imitate behaviour amongst the black community in the United States', put the government's imprimatur on police behaviour. The previous year the press and the police had discovered a 'frightening new strain of crime' and 'mugging' was added to 'Sus' as an offence on which the police could go on the offensive against West Indian youth.[42] The courts had already nodded their approval — by way of an exemplary twenty-year sentence passed on a 16-year-old 'mugger'. From then on, the lives of black youths in the cities of Britain were subject to increased police pressure. Their clubs were attacked on one pretext or another, their meeting places raided and their events — carnivals, bonfires, parties — blanketed by police presence. Black youths could not walk the streets outside the ghetto or hang around streets within it without courting arrest. And apart from individual arrests, whole communities were subjected to road blocks, stop and search and mass arrests. In Brixton in 1975 the para-military Special Patrol Group (SPG) cruised the streets in force, made arbitrary arrests and generally terrorised the community. In Lewisham the same year the SPG stopped 14,000 people on the streets and made 400 arrests. The pattern was repeated by similar police units in other parts of the country.[43]

The youth struck back and the community closed behind them at Brockwell Park fair in 1973, for instance, and at the Carib Club (1974) and in Chapeltown, Leeds, on bonfire night (1975), and finally exploded into direct confrontation, with bricks and bottles and burning of police cars, at the Notting Hill Carnival of 1976 — when 1,600 policemen took it on themselves to kill joy on the streets.

Clearly the politics of the stick had not paid off — or perhaps needed to be stepped up to be really effective. But by now a Labour government was in power and the emphasis shifted to social control.

Meanwhile, the struggles in the workplace were throwing up another community, a community of black class interests — linking the shop-floor battles of Pakistanis, Indians and West Indians, sometimes directly through roving strike committees, sometimes through black political organisations, while combating at the same time the racism of the trade unions, from within their ranks. Where they were not unionised, black workers first used the unions, who were rarely loth to increase their numbers, however black, to fight management for unionisation — and then took on the racism of the unions themselves. Unions, after all, were the organisations of their class and, however vital their struggles as blacks, to remain a people apart would be to set back the class struggle itself. They had to fight simultaneously as a people and as a class — as blacks and as workers — not by subsuming the race struggle to the class struggle but by deepening and broadening class struggle through its black and anti-colonial, anti-imperialist dimension.

The struggle against racism was a struggle for the class.

A series of strikes in the early 1970s in the textile and allied industries in the East Midlands and in various factories in London illustrate these developments. In May 1972 Pakistani workers in Crepe Sizes in Nottingham went on strike over working conditions, redundancies and pay. They composed the lowliest two-thirds of the workforce, were subjected to constant racial abuse by the white foreman and worked, without adequate safety precautions and toilet and canteen facilities, an eighty-four-hour week for £40.08. And yet five of their number had been made redundant — after the workers had joined the Transport and General Workers' Union (TGWU). There was no official support from the union, however, till a Solidarity Committee composed of the wives and families of the strikers and of other Asian workers, community organisations and the Nottingham-based BPFM forced the TGWU to act. In June the management capitulated, agreeing to union recognition and the re-instatement of the workers who had been made redundant.

The strike of Indian workers at Mansfield Hosiery Mills in Loughborough in October 1972 was for higher wages and against the denial of promotion to jobs reserved for whites. The white workers went along with the wages claim but not promotion, and the union, the National Union of Hosiery and Knitwear Workers, first prevaricated and then decided (after the strikers had occupied the union offices) to make the strike official, but not to call out the white workers. Once again, community associations, Asian workers at another company factory and political organisations like the BPFM and the BWM provided, in the Mansfield Hosiery Strike Committee, the basis for struggle.

So that when strikes by Asian workers at the Courtauld-owned Harwood Cash Lawn Mills in Mansfield and E.E. Jaffe and Malmic Lace in Nottingham broke out in the middle of 1973, the Mansfield Hosiery Strike Committee was at hand to give them support. More importantly — from a long-term view — the Strike Committee, pursuing its policy of pushing the trade union movement to fight racism not just in word but in deed, now called for a Conference of Trade Unions against Racialism. Accordingly, in June 1973 350 delegates from all the

the major unions and representatives from black community groups and black political organisations[44] came together at a conference in Digbeth Hall, Birmingham. From this emerged the Birmingham Conference Steering Committee, which in turn led to the setting up of the National Committee for Trade Unions against Racialism (NCTUAR).

Meanwhile, in the London area in June 1972, West Indian workers at Stanmore Engineering Works struck work demanding wage increases recommended by their union, the Amalgamated Union of Engineering Workers (AUEW). They went further — they staged a sit-in. But although the union was prepared to award strike pay, it was not prepared to bring its national weight to bear on the strike — by, for instance, getting workers and unions in the motor industry to 'black' products from Stanmore Engineering. The strikers were eventually removed by a court injunction and sacked.

Trade union racism showed up again a year later in the strike at Standard Telephone and Cables (New Southgate), a subsidiary of ITT, over promotion of West Indians to 'white-only jobs'. The craft unions, like the Metal Mechanics, remained stubbornly craft/race oriented. The Electrical Trades Union (ETU) opposed the strike as detrimental to its (white) members, into whose ranks blacks sought promotion. The local AUEW shop steward, though supporting the strikers, could not get the support of his union on a national basis. The NCTUAR called on the trade unions and trade unionists to back the workers in their official strike action against racial discrimination — and leafletted the annual Trades Union Congress (TUC) Conference at Blackpool. Once again, all the black political organisations, the London-based BUFP, BCC, BWCC and the BWM, along with the BWAC and the BPFM from the East Midlands, came to the aid of the strikers. The BWAC sent a cable to the Non-Aligned Conference in Algiers pointing out the international depredations of ITT. But all to no avail.

In November 1973, in a strike at Perivale Gütermann, a yarn factory in Southall, over the question of wages and productivity, Indian and Pakistani workers struck work and were sacked. The TGWU branch supported the strike but gave no strike pay till February the following year. Management tried to introduce their version of the Indo-Pakistan war into the factory, but failed to inflame communal passions. The workers once again turned to the communities for help and were assisted by Indian and Pakistani workers' associations, *gurdwaras* and local shops, who between them collected money and supplied the men on strike with free sugar, flour, oil and essential groceries. The TGWU, which, like most unions, had hitherto refused to cooperate with the government's restrictive Industrial Relations Act, now referred the case of the dismissed workers to the NIRC — which of course ruled against the men. The strike was defeated.

The apotheosis of racism, however, and therefore the resistance to it, was reached in 1974 at the strike in Imperial Typewriters (in Leicester), a subsidiary of the multinational Litton Industries. For here the white workers, management and unions worked hand in glove and were backed up by the violent presence of the National Front at the factory gates. Over a thousand of the 1,500 workforce were Asians, a large section of them women, most of them refugees from Uganda, and the strike itself arose from the usual practices of racial discrimination and exploitation. The TGWU refused to support the strikers with the hoary excuse that they had not followed correct negotiating procedures, and even prevailed on some of the Asian workers to remain at work by insisting that 'the tensions are between those Asians from the subcontinent and those from Africa'. By now, of course, there was virtually a standing conference of black strike committees in the Midlands and a network of community associations and groups plus a number of black political organisations, all of which came to the aid of the strikers. And money came in from, amongst others, the Southall IWA, Birmingham Sikh temple, a women's conference in Edinburgh, the Birmingham Anti-racist Committee and the European Workers' Action Committee. The strikers won, but the firm was closed down shortly afterwards by the multinational parent company.[45]

By the middle of the 1970s, the youth had begun to emerge into the vanguard of black struggle. And they brought to it not only the traditions of their elders but an experience of their own, which was implacable of racism and impervious to the blandishments of the state. The daily confrontations with the police, the battles of Brockwell Park and Chapeltown and Notting Hill and their encounters with the judicial set-up had established their hatred of the system. And they were now beginning to carve out a politics from the experiences of their own existence. Already by 1973, 'marginalised' young West Indians in the ghettos of Britain were being attracted to the popular politics of Rastafari. Bred in the 'gullies' of Jamaica, the Rastas were mortally opposed to consumer-capitalist society and saw in their own predicament the results of neo-colonial and imperialist intervention.[46] And in their locks and dress and music they signified their deadly opposition. They were the 'burning spear' of the new resistance. The police took note, the state also.

The Labour government's White Paper of September 1975 and the Race Relations Act that followed it in February 1976 spelt out between them the anxieties of the state. Having noted that 'about two out of every five of the coloured people in this country now were born here and the time is not far off when the majority of the coloured population will be British born', the government warned that it was 'vital to our well being as a society to tap these reservoirs of resilience, initiative and vigour in the racial minority groups and not to allow them to lie unused or to be deflected into negative protest on account of arbitrary and unfair discriminatory practices'. Hence the government would pass a Race Relations Act which would encompass whole areas of discrimination and vest the new Commission for Racial Equality (CRE), a merger of the CRC[47] and the RRB, with a few more powers to deal with it — and develop in the process a class of collaborators who would manage racism and its social and political fall-out. At the same time, it would hand out massive sums of money from its Urban Aid programme to key black self-help groups and so stamp out the breeding-grounds of resistance.[48]

The strategy worked in the short run. But even within a year, it was showing signs of failing in the long term. In September 1975 three West Indians (two of them youngsters), in the hope of financing black political groups that had refused to be corrupted by state benefice and of setting up black schools and self-help groups, held up the Spaghetti House, a restaurant in Knightbridge. At the end of a five-day siege, they were arrested and charged and received sentences from seventeen to twenty-one years. For Sir Robert Mark, Metropolitan Police Commissioner, 'the Spaghetti House case... was the most difficult and potentially explosive of all the various problems' he had to deal with in his career,[49] but it was also one in which his strategy to by-pass his political masters and go direct to the media for the legitimation of police practice had paid off. It was an entente that, given the endemic racism of the media and the police, operated naturally vis-à-vis the black population, but would now be extended to other areas of society and substitute legitimation for accountability.[50]

Among the Asians, too, it was the youth who were moving into the forefront of struggle. Like their Afro-Caribbean peers, they had been bred in a culture of racism and, like them, were impatient though not dismissive of the forms of struggle that their elders conducted. The fascist attacks in their community had gone on mounting, the police afforded no protection against them, condoned them, even, by refusing to recognise them as racially motivated. And the police themselves subjected the community to racial abuse, arbitrary arrest and 'fishing raids' for 'illegal immigrants'. And then, in June 1976, opposite IWA's Dominion Cinema, Southall, a symbol of Asian self-reliance and security, 18-year-old Gurdip Singh Chaggar was set upon by a gang of white youths and stabbed to death. (The motive, announced Sir Robert Mark, was not necessarily racial.)

A few months earlier, the government (Labour) had announced a Green Paper on Nationality (on the lines of the present Tory Act) which would 'rationalise' the law which they themselves had fouled up in 1968. (Of course they had, as was their wont, balanced it with an anti-discriminatory Race Relations Bill which was just then, March, going through parliament.) In April the NF staged a march through the black areas of Bradford under police protection, but were beaten back by the people of Manningham, Asians and Afro-Caribbeans, young and old. In May the press started a concerted campaign against immigration with the revelation that a homeless British Asian family expelled from Malawi was being housed in a four-star hotel at a cost of £600 per week to the British tax-payer. Later that month, Enoch

Powell announced that he had secret information from a 'suppressed' government report which said that bogus dependants and wives from India were making their way into the country. The press picked up Powell as Powell had picked up the press. And the attacks on the 'Asian invaders' became more intense through the days of May. On 4 June Chaggar was killed.[51]

The community was stunned. A meeting was held and the elders went about it in the time-honoured way, passing resolutions, making statements. The youth took over — marched to the police station, demanding redress, stoning a police van en route. The police arrested two of them. They sat down before the police station and refused to move — until their fellows were released. They were released. The following day the Southall Youth Movement (SYM) was born.

Various Asian youth movements sprang from this initiative — whenever and wherever there was need and in response to specific circumstances. But since these circumstances were invariably connected with fascist attacks and murders and/or police inability either to protect or apprehend (an inability so massive that it had taken a qualitative leap into connivance),[52] the youth movements tended to centre largely around the defence of their communities, and their organisations to reflect that purpose. (Their intervention in the campaigns against deportation would come later.) In the course of the next couple of years a number of youth organisations and defence committees sprang up, in London, Manchester, Leicester, Bradford, several of them in London alone — in Brick Lane after the murders of Altab Ali and Ishaque Ali, in Hackney after the murder of Michael Ferreira, in Newham after the murder of Akhtar Ali Baig. And, like the strike committees earlier, the youth groups moved around aiding and supporting each other — joining and working with West Indian youth groups in the process, sometimes on an organisational basis (SYM and Peoples Unite, Bradford Blacks and Bradford Asian Youth Movement), sometimes as individuals, often coalescing into political groups (Hackney Black People's Defence Organisation and Bradford's United Black Youth League).

At another level, political groups were consciously formed by Afro-Caribbeans, Asians and Africans who had been active in white left movements but had left them because they did not speak to the black experience. And they took on not only the black condition in Britain, but that of black peoples everywhere. They were anti-racist and anti-imperialist; and they were active in their communities. Their publications showed these concerns and helped further to politicise black people. *Samaj in'a Babylon*, in Urdu and English, and the group that produced the paper, came out of Chaggar's murder (June 1976), the Notting Hill riots (August 1976) and Soweto (June 1976). *Black Struggle* was its theoretical but accessible counterpart, *Mukti* its successor. Black Socialist Alliance (BSA) would comprehend them all for a while and shift the emphasis to campaigning material. Blacks Against State Harassment (BASH) would later address itself specifically to state racism. Other papers and journals and defence committee sheets and newsletters came and went, like their organisations, as the struggle rose and fell, moved and shifted, re-formed — but moving always in one direction: against the police, the government, racism. And the sheep/goat distinction that the state had hoped, by selective openings in higher education, to achieve, had broken down: the educated gave of their skills to the community and the community grounded them in the realities of political struggle.

Stop at Heathrow a minute, at the airport, as you are coming in or, if you are lucky, going out. Look around you, and you will see the division of labour that characterises the workforce of Britain. Cleaning and sweeping the (women's) lavatories, the halls, the stairways are Asian women from nearby Southall. Among the porters you will find a scattering of Asian and West Indian men. In the catering section, white women pack the food on the trays, while Asian women pack the same trays with cutlery (for £10 less per week). The menials in the kitchen are invariably Asian women — plus a few men, perhaps, for the heavier work.

And, of course, there is no question of promotion. Indeed they are lucky if the agency that employs them does not sack them and re-employ them at some other terminal, at the same wage if not a lower one. Their union, the TGWU, has been indifferent to their demands and in 1975, when 450 Asian workers walked out on their own initiative, the union declared the strike unofficial. The women managed, though, to elicit a few concessions on their own — and went back to work.[53]

The strike at Grunwick Film Processing plant in North London in August 1976 is, of course, more celebrated — not only because the Asian workers, most of them women from East Africa, sustained it in wet and snow and police harassment of pickets for over a year, but also because the whole force of the unions and of government appeared to be gathered at last on behalf of black workers. Not only were the strikers given strike pay by their union, but were also supported by the national unions — TGWU, TUC and UPW (Union of Post Office Workers) — and by local union branches, shop stewards committees, trade councils, the lot. And cabinet ministers appeared on the picket lines. The basic issue for the strikers was the question of racist exploitation with which union recognition was involved, but, in the course of accepting union support, they also accepted the union line that union recognition by management was really the basic issue, losing in the process the lasting support of the black community. Union recognition would not have of itself got the vast backing of the unions, let alone that of cabinet ministers — it had never happened before — but there was now a deal between the government and unions (the Social Contract) which in exchange for workers not striking ensured, through the Employment Protection Act, that employers did not prevent unionisation. And that put Grunwick in the middle of it.

As the strike dragged on into a year and the media and the management and its supporters threatened to involve more fundamental political issues such as the closed shop and the mass picket, the unions lost interest and left. In November 1977 four Asian strikers, two of them women, started a hunger strike outside the TUC headquarters. They were immediately suspended by their union and their strike pay withdrawn. Len Murray, General Secretary of the TUC, suggested that they take up their hunger strike at the factory gates and not outside his office.[54]

The lessons of the earlier strikes — that black workers needed to rally the community behind them and from that base force the unions to their side — had been temporarily unlearnt by workers who had not had the benefit of that tradition. On the other hand, the persistence of Asian women in going on the picket lines, month after month, against the pressure of their husbands and their fathers, the deception of the union and the attacks of the SPG — supported consistently by women's groups — had established the strength of the emerging black women's movement.

In 1977 the National Front, encouraged by their performance (in terms of the percentage of votes cast) in previous local elections, staged several marches through black city areas, with the police ensuring for them the freedoms of speech and assembly. They were closely attended by anti-racist groups — and black youth took the opportunity to stone both police and fascists alike.

In January 1978 Judge McKinnon ruled that Kingsley Read's pronouncement on Chaggar's death — 'One down, one million to go' — did not constitute incitement to racial hatred. 'In this England of ours,' the Judge observed, 'we are allowed to have our own views still, thank goodness, and long may it last.' Kingsley Read was the head of the fascist National Party.

In the same month, in the run-up to the local elections, itself a run-up to the general election of the following year, Margaret Thatcher assured the nation that her party would 'finally see an end to immigration', for 'this country might be rather swamped by people with a different culture'. Since primary immigration had ended with the 1971 Act, she was clearly referring to dependants. Shortly afterwards, the House of Commons (all-party) Select Committee on Race Relations and Immigration, sounding a similar note, went on to recommend 'new procedures to tighten up identity checks' and 'the consideration of a system of internal control on immigration'. The Tories promised to go further: they would 'improve' existing 'arrangements... to help those who are really anxious to leave this country'. The existing 'arrangements', such as the SPG and the IIIU, the immigration officials, and the Home Office, the courts and the media, were obviously not enough; the Tories would reverse policies of 'reverse discrimination' and amend the law on incitement to racial hatred, requiring it to prove 'an intent to offend'.

The media quickly tuned into Thatcher's warning about 'swamping'. The *Daily Mail*, in a series of articles on immigration, with headlines such as 'They've taken over my home town', gave real life stories of 'culture swamping'. A BBC television discussion programme on immigration afforded Enoch Powell enough latitude to enlarge on his theme of 'induced repatriation'. In the local elections that followed in May, Tory candidates reiterated and justified Tory proposals.

Hardly had the orchestration ceased than white fascist maggots began to crawl out of the decaying capitalist matter. Whole communities were terrorised. Three Asians were murdered in London within a period of three months, a shot-gun attack was mounted on West Indians in Wolverhampton, places of worship were desecrated and properties damaged and vandalised. And Tory-controlled local councils, mandated by their victory at the polls, set out to pursue Thatcherite policies in preparation for her victory.[55]

Emboldened by these events, but also wishing to show the country that they were the true party of fascism, the NF in April the following year requested permission from the local council to hold an election meeting at Southall Town Hall. Permission had been refused elsewhere and, even in Ealing, refused by the previous Labour-controlled council. Now, however, the Tory majority, after little prevarication, granted permission. Five thousand people demonstrated before the Ealing Town Hall the previous day, demanding that the meeting be called off, but to no avail. Instead the council flaunted the Union Jack, the NF's symbol, from the roof of the town hall. It was St George's day, a day celebrated by the NF. The Southall community planned a peaceful protest. 'But on the day, 2,756 police, including SPG units, with horses, dogs, vans, riot shields and a helicopter, were sent in to crush the protest' — and the whole town centre declared a 'sterile' area.[56] People were penned in, unable to get to the town hall or go back home — and began milling around. The police went berserk. Police vans were driven at crowds of people and when they scattered and ran, officers charged after them, hitting out at random. Blair Peach, a relentless anti-racist campaigner and teacher, was beaten to death and hundreds of others injured, many seriously. The offices of Peoples Unite (an Afro-Caribbean-Asian meeting centre) were vandalised by police in readiness for the Tory council to demolish them — years before the scheduled date. Asian newspapers recalled the Amritsar massacre of that other April in 1919.

The trials of the Southall 342 were held twenty-five miles away, far from the eye of the community, in Thatcher country. The magistrates rushed rapidly through the cases, convicting with abandon — 80 per cent in the first weeks of the trials — before the community could alert public opinion and the conviction rate was brought down to 50 per cent. The SPG officer who had bludgeoned Blair Peach to death remained unidentified and untried. The (Tory) government refused to hold an inquiry. The Home Secretary tut-tutted the SPG and, despite a massive public outcry against the unit (in which even the media was caught up), let it go back to its former devices. The Metropolitan Police Commissioner, Sir David McNee, summed it all up in an epigram: 'If you keep off the streets in London and behave yourselves, you won't have the SPG to worry about.' But Southall, Southall knew, would not lightly be invaded again, as the third of July was to prove.[57]

The Tory government of 1970, with its Immigration Act and Industrial Relations Act, its White Paper on Police-Immigrant Relations and other bits and pieces of policy, had begun the moves from the control of blacks at the gates to their control within. The Labour government had continued in the same vein, and not always by default: there was the Child Benefit law[58] and the Green Paper on Nationality — and, of course, there was their unwavering support for the police and their practices. The Tory government of 1979 now sought to perfect these measures, carry them to a logical conclusion, a final solution, within an overall attack on the working class and the welfare state in the framework of a law-and-order society. In articulating and clarifying the ideology of British racism in the run-up to the elections, Thatcher had established a climate in which officials in the health service, employment, education, housing, social and welfare services would, without benefit of edict, insist on passports and identity checks before affording a service to black citizens. Her Nationality Bill, by providing for various classes of (black) citizenship, would tend to regularise these practices.[59] Britain was effectively moving to a pass-law society.

The resistance of the black community went up a notch and, as so often before, threw up new types of struggle and new leaderships — this time in the form of the black women's movement, which would encompass all the struggles and add its own particular perspective to the resistance of the late 1970s. A few Afro-Caribbean women's groups had been in existence for over a decade, taking up issues that neither the white women's movement nor the black parties would concern themselves with. Asian women had begun to support their sisters through industrial strikes, on the Grunwick picket line, for instance, and outside Heathrow Airport. By 1978 black women's groups, Asian and Afro-Caribbean, had sprung up all over Britain and came together to form one powerful national body, the Organisation of Women of Asian and African Descent (OWAAD), with its paper *FOWAAD*. OWAAD would hold national conferences and work with other national black groups, whilst allowing its constituent groups autonomy of work in their communities. Through OWAAD, the Asian and Afro-Caribbean experiences and campaigns could cross-fertilise and develop particular lines of struggle that would benefit the whole black community. For they were taking up issues of discrimination against class, race and gender at once — in the face of harassments which, under the new Tory regimen, went deep into community life, into households, into children's welfare.

It was naturally Asian women's groups which first became aware of issues such as the discrimination in new Child Benefit provisions — since it was mainly their community's children which were kept out by immigration law. But they were soon joined in their campaigning by Afro-Caribbean women who were already exposing other state attacks on black family life. Together they worked on issues of black child care, black prisoners' rights, the enforced use of Depo-Provera and abortion law (without recourse to abortion, black women would be subjected increasingly to dangerous contraception methods such as the use of Depo-Provera). Asian women joined the campaigns against 'Sin-bins' (special 'adjustment units' which replaced ESN schooling for West Indian children), which the United Black Women's Action Group in North London had started. Brixton Black Women's Group launched the first Black Women's Centre (1979).

In fighting for educational and social and welfare services for the whole community, Asian and Afro-Caribbean women pinpointed the parallel histories of a common racism. In health care, for example, black women fought against the neglect of 'black disease'. Simultaneous campaigns were mounted in Brent against sickle-cell anaemia (affecting West Indians) and Vitamin D deficiencies causing rickets (affecting Asians). And issues such as forcible sterilisation, arising from the health services' obsession with black fertility, or the easy consignment of black women to mental hospitals, arising from its stereotyped understanding of 'black psyches in captivity', were fought by black women from both communities.

Asian women in *AWAZ* (Voice) and Southall Black Sisters (set up in the wake of 23 April 1979) continued to lead the protest against the virginity testing and X-raying of immigrants. Here they gave the lead not only to other women, but to long-established, male-dominated Asian organisations such as the IWA, which eventually joined them. And when Asian youth groups began to campaign in the community over specific immigration cases, it was the black women that helped keep the names of Anwar Ditta and Nasira Begum in the public consciousness.

Black women have also been active in working-class struggles, as in the strike of Asian women at Futters (March-May 1979) and at Chix (1979-80), worked in local community self-defence groups and combined in national black campaigns such as BSA and BASH. And from the richness of their struggles — at the factory gate, on the streets, in the home, at the schools, in the hospitals, at the courts — and from their joint initiative with IWA and BASH arose the first national black demonstration against state brutality (June 1979), when blacks, with the violence of virginity tests, the fascists and the SPG still fresh in their minds, marched in their outraged thousands through the heart of London.[60]

The loom of British racism had been perfected, the pattern set. The strands of resistance were meshed taut against the frame. The frame had to give. Instead, it was screwed still tighter in the unexplained death of Rasta 'Cartoon' Campbell in Brixton prison[61] (March 1980), the murder of Akhtar Ali Baig on the streets of Newham (July 1980) and the burning to death of thirteen young West Indians in a fire in New Cross (January 1981).

It was clear to the black community from the evidence of a black witness, if not the evidence of their whole history in Britain, that the fire had been started by the fascists. But even before the investigation was concluded, the police, with the aid of the press, had put it about that it was the work of a disaffected black party-goer, or a prank that went wrong or maybe an accident. Finally they 'proved' through forensic expertise that the fire had been self-inflicted one way or another. Just nine months earlier, the police had 'raided' a black meeting place in St Pauls Bristol, only to be beaten back by the youth and routed by the community. Now the community closed ranks again. From all over the country they gathered at meetings in New Cross. A day of action was planned. The Race Today Collective[62] took over its organisation. And on 2 March over 10,000 blacks downed tools and marched through the heart of London, past the halls of imperial finance, past the portals of the yellow press, past the Courts of Justice, past the proud shopping centre of consumer society, past Broadcasting House and into the anointed place of free speech — Speakers' Corner.

It had been, for its size and length and spread of time, a peaceful march. There had been a few skirmishes, a window or two broken and a few arrests made. But the banner head-lines in the people's press spoke of 'mob violence', 'blacks on the rampage', the invasion of privacy, the damage to property. The quality press mourned the breakdown of police/black relations, the frustration of the blacks, even at times white insensitivity to 'black problems' — and went back to sleep again. The Home Secretary muttered something about an inquiry into racial violence. White society ensconced itself in its goodness and thanked God for the British 'bobby'. And, heartened, the bobbies went back to baiting Brixton, the fascists to baiting Southall.

In April, Brixton exploded in rebellion, in July, Southall — for blacks, Afro-Caribbean and Asian alike, all distinction between police and fascist had faded — and in the days following, Liverpool, Manchester, Coventry, Huddersfield, Bradford,

Halifax, Blackburn, Preston, Birkenhead, Ellesmere Port, Chester, Stoke, Shrewsbury, Wolverhampton, Southampton, Newcastle, High Wycombe, Knaresborough, Leeds, Hull, Derby, Sheffield, Stockport, Nottingham, Leicester, Luton, Maidstone, Aldershot and Portsmouth, black and white — rebellion in slum city — for the deprived, the state was the police.

Nowhere have the youth, black and white, identified their problems with unemployment alone. That has been left to the social analysts of a past age. The youth know, viscerally, that there will be no work for them, ever, no call for their labour: it was not just a matter of the recession (the rich were doing all right), technology was taking over and the recession just gave 'them' the chance to get rid of the workers and bring in the robots. Society was changing, and they didn't need the secretary of the British Association for the Advancement of Science to tell them that it was 'a fundamental and irreversible change'. But they do not want to be pushed into artificial work schemes or institutionalised leisure or receive hand-outs from the enemy state. There is enough to go round and they want a part of it, a say in its giving. Or they will get it by thieving and 'loitering' and hustling — those things which pass for normalcy in a slum but threaten established society.

They are not the unemployed, but the never employed. They have not, like their parents, had jobs and lost them — and so become disciplined into a routine and a culture that preserves the status quo. They have not been organised into trade unions and had their politics disciplined by a labour aristocracy. They have not been on the marches of the dis-employed, so valiantly recalled by Labour from the hunger marches of the 1930s. Theirs is a different hunger — a hunger to retain the freedom, the life-style, the dignity which they have carved out from the stone of their lives.

The police are not an intrusion into that society but a threat, a foreign force, an army of occupation — the thick end of the authoritarian wedge, and in themselves so authoritarian as to make no difference between wedge and state.

That authoritarianism had been perfected in the colonies, in Ireland, in the fields of British racism, and, as it grew, it found ways to by-pass its political masters and become accountable to no one but itself — by obtaining legitimation for its actions from the silent majority through its cultivated liaison with the media.

It was once held that the British police were governed more by popular morality than by the letter of the law. They have now become the arbiters of that morality. There is no criticism of them they would brook, no area of society they do not pronounce on (with the shadow of force behind them). Look at the ferocity with which they attack their critics (even the parliamentary tribunes of the people),[63] their refusal to be accountable to elected local police authorities, their pronouncements on the jury system, the unemployed, homosexuality, etc.,[64] the press campaign mounted by their PRO, the Police Federation, for increased police powers (in various submissions to the Royal Commission on Criminal Procedure for instance) or the bon mots of their police chiefs[65] to understand how the police have moved from acccountability to legitimation.

But then a government which is not accountable to the people — a government which governs with the politics of the stick and the policies of a thousand cuts, which is anti-working class and anti-women and anti-youth — must have a police force that is accountable to it and not to the people. In turn, the government itself needs to be legitimated by an ideology of repression. And it is not merely that a free market economy requires a law and order state but that, even in its passing, it leaves only the option of a mixed economy with a corporate state maintained by surveillance. They are but two shades of the same authoritarianism, the one more modern than the other, but neither speaking to the birth of a new society that waits in the wings of the new industrial revolution.

Notes

1. Ruth Glass and Harold Pollins, *Newcomers*, London: Centre for Urban Studies and George Allen & Unwin, 1960.
2. D.R. Manley, 'The social structure of the Liverpool Negro community with special reference to the formations of formal associations', unpublished thesis (1958).
3. *Keys*, Vol. 3, no. 2, October-December 1925.
4. Edward Scobie, *Black Britannia*, Chicago: Johnson Publishing Co., 1972.

5. From it sprang *Link, Carib, Anglo-Caribbean News, Tropic, Flamingo, Daylight International, West Indies Observer, Magnet* and others.
6. 'Without integration, limitation is inexcusable; without limitation, integration is impossible.' Roy Hattersley, 1965.
7. That Fenner Brockway, a ceaseless campaigner for colonial freedom, had introduced a Private Member's anti-discrimination bill year after year after year from 1951 had, of course, made no impact on Labour consciousness.
8. *West Indies Observer*, Vol. 1, no. 19, 4 May 1963.
9. *West Indies Observer*, Vol. 1, no. 22, 15 June 1963.
10. *West Indies Observer*, No. 36, 18 January 1964.
11. Joseph A. Hunte, *Nigger Hunting in England*, London: West Indian Standing Conference, 1965.
12. Commonwealth Immigration Advisory Council, *Second Report*, London, 1964.
13. Department of Education Circular 7/65, London, 1965.
14. Paul Foot, 'The strike at Courtaulds, Preston', *IRR Newsletter* Supplement, July 1965.
15. Peter Marsh, *The Anatomy of a Strike*, London: Institute of Race Relations, 1967.
16. Raas, a Jamaican swear word, gave a West Indian flavour to Black Power.
17. I remember the time in South London when an old black woman was being jostled and pushed out of a bus queue. Michael went up and stood behind her, an ill-concealed machete in his hand — and the line of lily-white queuers vanished before her — and she entered the bus like royalty.
18. The National Federation of Pakistani Associations was formed in 1963.
19. Times News Team, *The Black Man in Search of Power*, London: Nelson, 1968.
20. A. Sivanandan, *Race and resistance: the IRR story*, London: IRR, 1974. See also Jenny Bourne and A. Sivanandan, 'Cheerleaders and ombudsmen: the sociology of race relations in Britain', *Race & Class*, Vol. XXI, no. 4, 1980.
21. In real life and real struggle, the economic, the political and the ideological move in concert, with sometimes one and sometimes the other striking the dominant note — but orchestrated, always, by the mode of production. It is only the marxist textualists who are preoccupied with 'determinisms', economic and otherwise.
22. Dilip Hiro, *Black British, White British*, Harmondsworth: Penguin, 1971.
23. *The Times* (24 October 1967) quoted in *IRR Newsletter*, December 1967.
24. *IRR Newsletter*, December 1967.

25. *Ibid.*
26. Such as the Naxalites, Adivasis, Dalit Panthers in India, and the Pakhtun, Sindhi and Baluchi oppressed people's movements in Pakistan. In 1974 organisations of untouchables in Britain came together at the (new) IRR to organise an International Conference on Untouchability (which for financial reasons never got off the ground).
27. Quoted in E.J.B. Rose et al, *Colour and Citizenship*. London: Oxford University Press for IRR, 1969.
28. There were in fact about 66,000 at this time who were entitled to settle in Britain.
29. Jagmohan Joshi, quoted in C. Karadia, 'The BPA', *IRR Newsletter*, June 1968.
30. A reference to Powell's Birmingham speech (April 1968) in which he said: 'As I look ahead, I am filled with foreboding. Like the Roman, I seem to see "the River Tiber foaming with much blood".'
31. The Bill proposed to give immigrants who were refused entry the right of appeal to a tribunal.
32. A. Sivanandan, 'Imperialism and disorganic development in the silicon age', see pp. 143-61 below.
33. JCWI was set up in 1967 as a one-man welfare service for incoming dependants at Heathrow Airport, but later burgeoned into a case-work and campaigning organisation.
34. According to a letter in the *Guardian* of 10 September 1981 from members of Jawaharlal Nehru University (Delhi), there are still '20,000 people of Indian origin from East Africa... waiting in India for their entry vouchers to the UK'.
35. See, for example, Robert Moore and Tina Wallace, *Slamming the Door*, London: Martin Robertson, 1975.
36. See 'Notes and documents' in *Race & Class*, special issue, 'Rebellion and repression: Britain '81', Vol. XXIII, nos. 2/3, 1981/2.
37. John La Rose had been Executive of the Federated Workers' Trade Union in Trinidad and Tobago.
38. It was one of the founders of this school, Tony Munro, who was later to be involved in the Knightbridge Spaghetti House siege.
39. The BWM (Black Workers' Movement) was the new name the Black Panthers took in the early 1970s.
40. Louis Kushnick, 'Black Power and the media', *Race Today*, November 1970.
41. See Institute of Race Relations, *Police against black people*, London, 1979, and various issues of *Race Today* for the important trials of this period.
42. Stuart Hall et al, *Policing the Crisis*, London: Macmillan, 1978.
43. Institute of Race Relations, *Police against black people*, *op. cit.*
44. These included representatives from Indian, Pakistani and West Indian associations and from black political organisations such as the

BPFM, BUFP, BCC and the Black Workers' Co-ordinating Committee, etc.

45. For all the above strikes, see various issues of the *BPFM Weekly*, (later *Uhuru*), the *BWAC Weekly Review, Black Voice* and *Race Today*.

46. See Colin Prescod, 'The people's cause in the Caribbean', *Race & Class*, Vol. XVII, no. 1, 1975; Horace Campbell, 'Rastafari: culture of resistance', *Race & Class*, Vol. XXII, no. 1, 1980 and Paul Gilroy, 'You can't fool the youths', in *Race & Class*, Vol. XXIII, nos. 2/3, 1981/2.

47. The Community Relations Commission (CRC) emerged as the successor to the NCCI in the Race Relations Act of 1968.

48. A. Sivanandan, 'Race, class and the state: the black experience in Britain', see pp. 101-26 below.

49. Sir Robert Mark, *In the Office of Constable*, London: Collins, 1978.

50. See Tony Bunyan, *The Political Police in Britain*, London: Julian Friedmann, 1976, and S. Chibnall, *Law and Order News*, London: Tavistock, 1977.

51. 'Race and the press' in *Race & Class*, Vol. XVII, no. 1, Summer 1976.

52. IRR, *Police against black people*, *op. cit.*

53. Campaign Against Racism and Fascism/Southall Rights, *Southall: the birth of a black community*, London: Institute of Race Relations, 1981.

54. See 'Grunwick', pp. 126-31 below, and 'UK commentary' in *Race & Class*. Vol. XIX, no. 3, 1978.

55. A. Sivanandan, 'From immigration control to "induced repatriation" ', see pp. 131-40 below.

56. Campaign Against Racism and Fascism/Southall Rights, *op. cit.*

57. 3 July 1981 — the day on which Asian youths burnt down a public house at which a racist pop group and its skinhead fans had gathered. This event was part of the 1981 uprisings.

58. Tax relief in respect of dependent children was replaced by child benefit paid to wives — but those with children abroad were not entitled to it, even if they were supporting them.

59. See 'Notes and documents' in *Race & Class*, Vol. XXIII, nos. 2/3, 1981/2.

60. Symbolically, the man who had initiated so many of the black working-class and community movements of the early years and clarified for us all the lines of race/class struggle, Jagmohan Joshi, died on the march, of a heart attack.

61. Campbell was arrested on 1 March 1980 on charges he claimed were false. From 10 March he refused food and drink. On the 26th he was force-fed. On the 31st he was found dead in his cell. Steve Thompson, who was also a Rasta, had a year earlier been forcibly shorn of his locks and following his protest sent to Rampton Mental Hospital. A Home Office circular denying recognition of Rastafarianism as a religion in prisons had been issued in 1976 (60/76). In 1977, a white sociologist showed in a Cranfield Police Study how the Rastas were terrorising the police of Handsworth (Birmingham). In April 1981, the Home Office confirmed its circular. In June, 'Tubby' Jeffers collapsed in prison from refusing food that violated his Rasta beliefs.

62. The Race Today Collective emerged from the radicalisation of the Institute of Race Relations (1969-72) as an independent black journal and had grown, under Darcus Howe and John La Rose, into an activist collective.

63. For example, Greater Manchester Chief Constable James Anderton referred to police critics 'as creepy and dangerous minorities... who are obviously using the protection imparted by our very constitution in order first to undermine it and then eventually to displace it' (September 1980).

64. Juries, opined Sir Robert Mark, 'perform the duty rarely, know little of the law, are occasionally stupid, prejudiced, barely literate and often incapable of applying the law as public opinion is led to suppose they do' (*Observer*, 16 March 1975).

65. Declaring that 'prejudice is a state of mind brought about by experience', Detective Superintendent Holland identified long-haired, unshaven youths as the ones likely to have cannabis and West Indians hanging around in jeans and T-shirts as likely 'muggers' (*Guardian*, 14 September 1981).

Part Two

Black Power and Black Culture

The movement for Black Power, which began in the US in the mid 1960s was to alter profoundly our very perceptions of what political struggle was, and how it needed to be waged. It influenced radically, both in content and form, the movements of personal and political liberation that followed in its wake, including the women's movement. It was the catalyst which showed up the essential unity of the struggles against white power and privilege — whether in the US itself, in Britain, in Southern Africa, or in the former colonies of the Caribbean. Through it, black became a political colour with which other Third World activists and radicals could identify — the Dalit Panthers from among the untouchables of India took their name from the Black Panthers of the US. The first piece in this section, 'Black power: the politics of existence' (published in February 1971), presages some of the themes that are taken up more fully a decade later, in the first essay in this book, 'From Resistance to Rebellion'.

Black Power threw up, from within its own ranks, its own thinkers and activists, its own spokesmen and women. Some of these came to symbolise and epitomise the movement they were part of: George Jackson and Angela Davis, Huey Newton and Bobby Seale of the Black Panther Party. Other individuals have been almost forgotten — Jonathan, George's younger brother — while the type of publicity that surrounds Muhammad Ali today obscures his earlier importance. Paul Robeson is certainly not forgotten, yet his stature and importance for the black movement has rarely been truly recognised.

The final essay in this section, written in 1972, takes up the theme of struggle against white domination, this time in the sphere of culture; the struggle to decolonise the mind, and lay the foundation for a new culture, new values. In doing so, it charts the journey of the black intellectual from race to class, from 'taking conscience of himself' to coming to consciousness of class.

Black Power: The Politics of Existence*

'Black Power' is a political metaphor — comprehending, at once, the history, condition, manifesto and programme of the black people of America. Powerlessness, the antonym of power, has characterised their history and accounted for their present condition. To transform that condition is needed the power to determine their own lives and destiny — at all levels. The strategy, the programme to achieve such power entails their solidarity as a people. This in turn implies the concepts of self-help, pride, and an indigenous culture.

A metaphor indeed, but also, in the terse, explosive precision of its language, a resounding call to arms. And that is how it broke from the lips of the marchers on that hot trek from Hernando to Jackson in June 1966. But the mood that precipitated the slogan that heralded the movement had been engendered over centuries of oppression. The grief and rage and submission of those years had often burst into insurrection. But, when in December 1955, in Montgomery, Alabama, tired old Rosa Parks, ordered to give up the white seat she had occupied in the bus, answered simply 'No', her reply was echoed by a young pastor and his congregation. The massive bus boycott that followed — forty-two thousand black men and women walked to and from work for 381 days — was the beginning of non-violent mass action. A rebel spirit of deliberate no-saying had seized the black. He had stood up — and walked — in a sort of incandescent beauty that people seldom attain in the mass except in their quest for themselves. As yet, it would appear, he had not asked to be counted — in the fullness of his being. He had merely stood up, with his fellows, against the injustices of his immediate oppression. It was a prelude to sensibility.

Over the years that followed — in the course of the freedom rides, the sit-ins, the voter registration drives, the marches, the petitions, the demonstrations and the resulting beat-ups, killings, and hand-outs — the black man came to a consciousness of himself as he never had before. No longer would he accept the

---

* *Politics and Society*, vol. 1, no. 2, February 1971.

white man's image of him as 'the quintessence of evil', no longer would he set store by white promises. For, he had seen four little black children blown to bits as they knelt to pray in a church in Birmingham. He had watched the manic, hate-filled faces of white parents shrieking, 'Lynch her, lynch her — drag her over to this tree, let's take care of the nigger' as a fifteen-year-old black girl was turned away from Little Rock Central High School and into a white mob by a National Guardsman. And it was he who had stood in the rain and had sung 'The Star-Spangled Banner' as he was beaten down and arrested. All along he had played the game according to the white man's rules, only to find him changing the rules so as to keep on winning. He had, in a word, held up to the white man the values he so ardently professed and found in him not the faintest sign of recognition.

And then 'they' shot James Meredith down — in the back — as he made his lone 'pilgrimage against fear' down the highway of his native Mississippi. As Martin Luther King and Stokely Carmichael and Floyd McKissick gathered the civil rights workers together to resume the march that Meredith had begun, the black participant began to look askance at his white comrade. This self-assured young white who went back to his safe white community at nightfall — what, they asked each other, had he to do with the poor, uncertain, fear-ridden black? Why was the black always subject to white guidance, white help, white patronage of one sort or another? Surely he must 'do his own thing', he must go it alone. Why too, non-violence when it was not even safe for a black merely to walk his country? And why 'we shall overcome some*day*', why not 'freedom *now*'? The murmur rose to a crescendo and the outraged cry of 'Black Power' burst the heedless land.

That is one way of putting it — a symbolic way, perhaps, but none the less exact for all that. For, a mere narration of facts, of events, does not add up to the truth. A metaphor, besides, needs to be understood in terms of the spirit that informs it.

Intimated in the events and mood leading up to the cry of 'Black Power' were the philosophy and direction that black politics would take in the future. A year later Carmichael and Hamilton made them explicit in their book: *Black Power: the politics of liberation in America*. 'The concept of Black Power', they pointed out, 'rests on a fundamental premise: before a group can enter the open society, it must first close its ranks. By this we mean that group solidarity is necessary before a group can operate effectively in a pluralistic society.'[1] They pointed out that the Irish had created an urban political machine, the Italians had made a business of crime, the Jews had retained their religio-cultural heritage and, whatever the intitial resistance to these groups may have been, they had finally achieved their place in the larger society — on their own terms. The 'integration' that was offered the black, however, spelt further deracination for himself and the loss of expertise to the community. It contributed nothing to alleviate the condition of the black masses except 'the token rewards that an affluent society was perfectly willing to give.'[2] And by holding out the belief that moving into a white neighbourhood or into a white school were criteria of excellence, integration reinforced the myth of white superiority. In effect, 'Integration as traditionally articulated would abolish the black community... and not the dependent colonial status that had been inflicted upon it.'[3]

Certain basic themes of current black ideology began to emerge from Carmichael and Hamilton's analysis of American society. The measure of equality meted out to a people in such a society would depend on their strength as a group. The black American had hitherto 'bargained' from a position of weakness, due to his lack of racial and political consciousness. Pride in himself and his heritage and control of the institutions that governed his immediate life were, therefore, the prerequisites of the group persona that he would have to develop before he entered the 'open society'. In the process, he might well have to 'revamp' the existing structure of American society.

From the pride and identity argument arose the cultural nationalist sect of Imamu Ameer Baraka (LeRoi Jones) and Maulana Ron Karenga; from the corollary — the thesis of a revamped society — grew the revolutionary nationalist branch of Huey Newton and the Black Panther Party. The first tacitly acknowledges the concept of a pluralist society and hopes, within it, to find the power that will give the Afro-American a choice of opting out of or into white society; the latter recognises that racism is a symptom of the malaise in capitalist society and therefore society will itself have to be restructured — and the black man is the obvious historical agent of that process.

Cultural nationalism takes for its text Ron Karenga's tenet that 'culture... gives identity, purpose and direction. It tells us who we are, what we must do, and how we can do it.'[4] But culture itself is based on a value system — black culture on a black value system — and that could only be African — 'because we are Africans, no matter that we have been trapped in the West these few hundred years.'[5] But however African their orientation, the cultural nationalists place no emphasis on the back-to-Africa philosophy of Marcus Garvey or the separatist beliefs of the Black Muslims. They stress, instead, the cultural traditions of Africa — African dress, hair style, names, language (Swahili), religion (Islam), and the African sense of community. Spirit House, a three-storey tenement in Newark presided over by the poet, novelist, and dramatist, Ameer Baraka (LeRoi Jones), is a focal point of cultural activity. It is here that the African Free School imparts to black children and the Black Community Development and Defence to black adults 'a value system' which will make them 'move, instinctively, to do what is best'.[6] The emphasis is on the development of personality as opposed to individuality — for 'any real individualism would recognise and indeed respect the human personality, and, thereby, free mankind'.[7] Politically, the people are organised along concrete, ad hoc issues geared to building up their own political machinery. Thus in Newark earlier this year, it was the organisation and thrust of the cultural nationalists that were instrumental in securing the election of a black candidate as mayor. Business enterprises such as The First and Last Store are run on cooperative lines to serve the needs of the black community. Black art, first fostered in Baraka's Black Arts Repertory School in Harlem, now flourishes in Spirit House and in similar community projects in the black urban areas across the nation.

The revolutionary nationalists, on the other hand, recognise that no effective change can take place in the lives of the black people until the capitalist system itself is overthrown. But at the same time black people constituted a colony within the mother country. Their oppression was therefore twofold: as a class and as a nation. Class oppression could only be overcome by the broad masses of the people over a period of political time. Racial or national oppression, however, demanded immediate redress by the national minority. And for this purpose, the black community should be organised to defend itself against and, in the process, destroy the armies of institutional racism. It was only through the direct action of the victims themselves that they would achieve an end to victimisation. If the police patrolled the community, the community would 'patrol' the police. If the children went hungry, the people would hold up the black capitalists to ransom and provide breakfast for the children. If they wanted a pedestrian crossing at a dangerous intersection where so many children had been run over, the community would not go on petitioning a deaf government, but set up a traffic signal themselves — and keep it there, with a shotgun if necessary. All that the leaders could do was to take on themselves the brunt of oppression. Through precept and example, self-discipline and sacrifice of life, through their good works in the community, they would educate and politicise their people.

Of course, black culture and black identity are essential to the life of the community. But a black American is not just a displaced African. Africa determines his identity no more than America does. They are both his history, however: one tells him whence he came, the other where he is at — two strands of the same consciousness. To accept the one to the exclusion of the other is to submit once again to the type of 'double consciousness' that DuBois spoke of in another context: 'a sense of always looking at oneself through the eyes of others'.[8] Black culture, while recognising its African genesis, needs to embrace the American experience and squeeze out of it a culture and an identity uniquely Afro-American. It needs to be coherent and dynamic at the same time. 'To create oneself in terms of one's culture and to reshape society in terms of that creation are part and parcel of the same process. Becoming and doing belong to the same continuum.'[9] By sequestering itself from meaningful social and political action over a wider area, black culture runs the risk of falling prey to black capitalism which is no less dangerous than the white variety. To move against the system even as one discovers one's identity is to prevent such a contingency.

The nationalism of the black revolutionary, therefore, is outward-looking — in terms of the oppressed peoples both of his own country and of the Third World. No one would be free until all men are free. The business of free men is to relate to the freedom of others — and therein find 'the universality inherent in the human condition'.[10]

Although both sects of the black nationalist movements have their counterparts in Britain, the conditions that deposited the 'coloured' people on these shores have made for somewhat different orientations. Admittedly the common experience of slavery makes for a common bond between Afro-Caribbean and Afro-American alike. But, whereas the West Indian has moved on from a colonial status to independence (or more accurately to a neo-colonial one), the black man's relationship to white America is essentially colonial in nature. More significantly the Afro-American, because he, unlike the Afro-Caribbean, has no country, no 'land base', of his own, is locked in mortal embrace with his colonial master and must perish or die with him. The West Indian in Britain, however, can in the final analysis escape his debased status in this country by returning to his own. But in view of the massive and ongoing exploitation of his own country, his choice is between psychological servitude here and economic servitude back home — which is no choice at all. He is therefore faced with the dual problem of overcoming injustice and discrimination in this country and overthrowing the neo-colonial setup in his own.

The radical black of Britain sees both problems as dynamics of the one continuum. His battle is waged on both fronts simultaneously and because he is a product of both colonialism and slavery he is the common denominator in the struggles of Afro-America and the Third World alike. He is the link that connects the enslaved peoples and the colonised, the blacks of America and the peoples of Afro-Asia.

It is this link which, in Britain, accounts for the unity, however temporal and ad hoc, between the Asian and West

Indian groups. The Universal Coloured Peoples' Association was the first flower of this burgeoning unity. The Black Unity and Freedom Party, which grew out of a withered Universal Coloured Peoples' Association, is another. It is possible that this organisation, too, will die. But others will take its place — for, the line of growth is clear: unified militant action and the relentless demand for justice are replacing the begging-bowl syndrome of the black liberal era.

But this comparative militancy is only a prelude to the revolutionary Black Panther oriented struggles that will be taken up by 'the second generation'. For these are youngsters who will not have known any experience but with the British, and it threatens to be an experience akin to that of the blacks in America. It is they who will more closely approximate 'the colony within the mother country' status of their American counterparts. And it is they who will take up the same solutions. They will have no country of the mind to return to. They are here and now and will take what British society owes them — as fully fledged British citizens — and will not give.

This applies, too, to the 'second generation' Asian — for he too is no less the product of this society, and his experience of second class citizenship is no different from that of the West Indian's. His language, his customs, his social orientation which once were Indian or Pakistani are now as wholly British as those of his Caribbean neighbour. Black to him is no less the colour of oppression than to the West Indian — and black power is no less the answer to his ills.

In the doctrine of the pluralist society as envisaged by Roy Jenkins,[11] Britain finds temporary refuge from the problems of a racially mixed society. The 'first generation' Asians are already a people, a group, in their own right. Their life-styles have already been formed in the countries of their birth. And they can, as a group, be slotted into a pluralist structure. The 'first generation' West Indian, on the other hand, is only now beginning to discover that he is not British after all. He had been weaned on mother country mythology, had come home to see his queen, to play his and their game of cricket only to be rejected without ceremony. He was not white after all, he was not British. And so he sets about looking for himself and, out of his blackness, squeezes out an identity which is sheerly his own. Black pride, black culture, black self-determination, black capitalism — he has heard the cry abroad. Now it is his own. From the Spirit House of Ameer Baraka rises the Black House of Abdul Malik. He finds solidarity in his group, he is become his own person. But will he enter the pluralist society?

Or will he move on, through his growing political consciousness, to a point where, along with his denigrated black British children, he challenges the very structure of this society? Or will he return home to create a revolutionary situation there — as some have already attempted to do in Trinidad?

The answers are not clear, but the trend is unmistakable. And given that the 'contagion' of black consciousness grows much faster than white recompense, the auguries for a truly multi-racial society in this country are bleak indeed.

What the Black Power movements of Britain and America and comparable movements in the Third World indicate is the end of white hegemony. For almost half a millenium, Europe had ravaged the countries of Africa, Asia and America (North and South), imposed her religions and cultures on their peoples, and committed them to one sort of bondage or another. In the wake of that 'civilising mission', 'one thousand five hundred million natives'[12] lay torn and rootless. For a while the 'native' had allowed himself to be guided by the bright white lights of Europe's capitals — a beacon in the world of his dark savagery. For a while he had believed in the white world's 'narcissistic dialogue'[13] of eternal values. He had accepted that he himself represented 'not only the absence of values but also the negation of values... the corrosive element disfiguring all that has to do with beauty or morality'.[14] And he had learned to keep his distance, know his place, he had moved only at the white man's behest — and only so far as the white man had allowed him. Had he so much as quickened his step or shifted from his allotted place, he had been brought back by force — which, after all, was the only language he understood. At the same time, he had looked abroad and seen, with growing envy, that the white man's world was a well-fed world, free, healthy, full of good things, of laughter, of children growing straight and strong, while his benighted world was stricken with hunger and disease, and his children wizened at birth. Violated in every aspect of his being, he had 'life to fear rather more than death'.[15]

Gradually, it dawned on the 'native' that white values were meaningless and white promises ineffectual. The white man, it was obvious, did not believe in them himself. His ideology, as Sartre has observed, was 'nothing but an ideology of lies, a perfect justification for pillage, its honeyed words, its affectation of sensibility... only alibis for aggression'.[16] It was futile to expect that the 'native' would be handed his freedom on a platter. He must, instead, wrest it from his oppressor — with force, 'the only language he understood'. Violence lay like an incubus on his mind: violence would release him from inaction and despair, violence would regenerate him and make him man.

The sequence of action and reaction, of violence and counter-violence, is clear: the whites act, the blacks react — and in terms of history, there is no doubt as to where the sequence begins. To argue, in the circumstances, that violence is a matter of choice, that it is self-defeating, is to be impervious to the fact that for some of the oppressed peoples of the world — for the black man in America's cities, certainly — 'violence... is not a matter of choice... but a symptom of the fact that there is no other'.[17] That anyone's range of choice should be so reduced to this one inescapable condition is itself the measure of the violence that has been visited on him. And violence in our time does not need to be overt and obtrusive to be recognised as violence — for poverty is violence, and racism; and the coincidence of poverty and racism is a violence beyond endurance.

To argue, too, that Black Power in its reaction to racism is itself racist is to overlook the fact that racial prejudice is essentially the white man's problem. The black man is concerned merely to achieve his humanity. What keeps him from this achievement is white oppression. The need to oppress, the primitive notion of racial superiority, is the white man's burden. It is he who must choose to lose it. The black man, again, has no choice.

To put it differently, white racism is at one level a matter of choice, at another a matter of privilege, but at all levels an exercise in oppression. White racism incurs, somewhere down the line, the denial of human dignity; black 'racism' envisages the negation of that denial. It is 'the rhetoric of abstracted liberalism'[18] which accords them equal weight.

And the liberal, fearful of the backlash people, points out that 'Black Power' itself is an offensive, sympathy-losing phrase. 'Coloured Power', perhaps, or 'Negro Power' would have been so much more palatable to the white power structure and less disturbing of the white psyche. But this again is the white man's problem — for, the connotations of 'Black', created by the white man himself are so frightening, so evil, so primordial that to associate it with power as well is to invoke the nightmare world of divine retribution, of Judgment Day.

Notes

1. Stokely Carmichael and Charles V. Hamilton, *Black Power: The Politics of Liberation in America*, London: Cape, 1968.
2. *Ibid.*
3. *Ibid.*
4. Ron Karenga, *The Quotable Karenga*, ed. Clyde Halisi and James Mtume, Los Angeles: *US* Organization, 1967.
5. Ameer Baraka, 'A Black Value System', in *Black Scholar* 1, no. 1, November 1969, pp. 54-60.
6. David Llorens, 'Ameer (LeRoi Jones) Baraka', *Ebony* 24, no. 10, August 1969, pp. 75-83.
7. *Ibid.*
8. W.E.B. DuBois, *A Reading*, ed. Meyer Weinberg, London: Harper & Row, 1970.
9. A. Sivanandan, 'Culture and Identity', *Liberator* 10, no. 6, June 1970, p. 11.
10. Frantz Fanon, *Black Skin: White Masks*, London: Macgibbon & Kee, 1968.
11. On 23 May 1966, Roy Jenkins, then Labour Home Secretary, defined integration as 'not a flattening process of assimilation but as equal opportunity accompanied by cultural diversity, in an atmosphere of mutual tolerance'. Cited in E.J.B. Rose and associates, *Colour and Citizenship: A Report on British Race Relations*, London: Oxford University Press, 1969, p. 25.
12. Jean-Paul Sartre, 'Preface' to Frantz Fanon's *The Wretched of the Earth*, London: Macgibbon & Kee, 1965.
13. Frantz Fanon, *The Wretched of the Earth*, *op. cit.*
14. *Ibid.*
15. Jean-Paul Sartre, *op. cit.*
16. *Ibid.*
17. A. Sivanandan, 'Race: The Revolutionary Experience', *Race Today* 1, no. 4, August 1969, pp. 108-109.
18. C. Wright Mills, *The Marxists*, Harmondsworth: Pelican, 1963.

Huey Newton and the Black Renascence*

They sold the books to buy the guns to do the things the book had told them to do — the red book. Poetry in action — that is one sort of beginning of the Black Panther Party.

Or one could begin elsewhere and say that on a day in the fall of 1966, in a room in the offices of the poverty programme in Oakland, California, two black men sat down to write the beliefs and aims of their people. The ten-point manifesto *cum* credo which resulted from that meeting was not particularly original or radical. The writers, Huey Newton and Bobby Seale, were themselves aware of the fact when they wrote that they were merely reiterating 'what black people have been voicing all along over 100 years since the Emancipation Proclamation and even before that'. Basically, 'Huey was talking about some full employment, some decent housing, some education, about stopping those pigs from brutalising us and murdering us'. Basically, too, each of the party's aims was tied to the needs, beliefs and idiom of the 'lumpen proletariat': the document might have been written by any one of the denizens of the ghetto — so simple, direct and explicit it was.

That same simplicity and directness manifested itself in the way Huey Newton and his Black Panther Party for Self-Defence went about achieving their aims. And it is in that manifestation that is to be found the initial and sustaining revolutionary content of the party. The more formal revolutionary concepts, which were to bring them alongside Marx and Lenin and Mao and Kim Il Sung and thereby validate them for oppressed peoples everywhere, would emerge later. But for now they were concerned with achieving through the direct action of the victims themselves an end to their victimisation. If the police patrolled the community, you patrolled the police. If they pulled a gun on you, you pulled a gun on them — in self-defence. If the law was invoked to brutalise you, you used the law to keep yourself from being brutalised. If the children were hungry, you held up the

---

* Review of Bobby Seale, *Seize the Time: The Story of the Black Panther Party and Huey P. Newton*. London: Hutchinson 1970, in *Race Today*, vol. II, no. 12, December 1970.

black capitalists in your area to ransom — by boycotting their businesses, by publicising their exploitation of the black community — and provided breakfast for the children. If you wanted a pedestrian crossing at the intersection where so many children had been run over, you did not keep petitioning a deaf government. You put up a traffic signal yourself, as Huey Newton did, and kept it there, with a shot-gun if necessary, as Huey Newton did.

But Newton and Seale and Cleaver and Hilliard and Huggins (Erica) were leaders only in that it was they who practised most assiduously what they preached, they who took on themselves the brunt of oppression, they who went into prison and exile and they who died in the cause of their people. By precept and example, self-discipline and sacrifice of life, by their good works in the community, they educated and politicised their people. And that is revolution. To go underground at a time of savage repression would have been the sane thing to do — to continue the fight guerilla-fashion. But they would then have been lost to public view, written off, been erased from the minds of their people, or have come to lodge there only as a legend. To stay overground and to act out their principles and their faith for all the world to see — that is revolution.

Theory is meaningless to the poor and the deprived. They do not know, they do not want to know, about Marx and Lenin. Fanon is closer to them. But it is Malcolm X, the neighbour boy, whom they really 'dig'. He has relevance to them. He speaks directly to their experience. And the Panthers speak directly to their experience. But the Panthers know that oppression everywhere is woven from the same fabric. And so they listen to other voices, other minorities, revolutionaries of other lands — they relate to them, enter into ad hoc coalitions, spark off comparable movements among the poor whites, the Chicanos, the Puerto Ricans and set all humanity afire.

But their eternal base is in their own community. And there, if we would but only look, a new 'religion' is being born, a new messiah is come, not a single but a collective messiah — new testaments are being written, values hewn out anew, a new man is being born. From the arid wastes of our prisons and our ghettoes come the sound and the smell of man creating himself —

out of the rubble of our civilisation, out of the wilderness of unlove. The prophets are among us, and the preachers and the poets — as it was in the year 00 — if we would but only hear.

The Passing of the King*

There is no set-back in history except that we make it so. Tonight the black world weeps that their king has passed away. But tomorrow and tomorrow and tomorrow... every black man will have become his own king — for that is the legacy that Muhammad Ali leaves us even as he leaves the ring.

The white world had willed that the king should die. But it took the might of the most powerful, the most designing judicial system in the world to bring the king down. Joe Frazier was the unreckoning tool of that design.

Heavyweight boxing had, until the advent of Muhammad Ali, come to be associated with brute force. If it once had been the province of elegant gladiators like Gentleman Jim Corbett, it was as the sport of white men. But as the black man began to claim the game more consistently, the game itself became tainted with the stereotype image of the 'nigger'. It was a thing for brutes — hefty, slow-moving, slow-thinking sub-humans — a blood sport from which the white man would gather profit and pleasure at no great cost to himself. It was satisfying too, to his psyche, if he could gather the myriad frustrations that in his daily life he visited on niggers in general and embody them in a single super-nigger. Two super-niggers would be doubly cathartic — and the more explosive they were, the more orgasmic his release.

At worst it was a game of make-believe, a confidence trick wherein the Thing, by being allowed to become a Super-Thing, believed it passed for man. Or so it would have seemed — until the black renascence of the sixties. Even the civil rights struggle had only served to cordon off the black athlete in a bantustan of sport. It was left to Malcolm X and the black power movement

* On Muhammad Ali, in *Race Today*, vol. III, no. 4, April 1971.

to threaten the total release of the Negro.

Muhammed Ali is the finest epitome of that release. And it is this that bugs white society so. He is not just a prize-fighter, he is not even one man. He is many men in one — and all of them black. Out of the very blackness, which white society decrees as evil, he squeezes out an image of himself and of his people which is both peerless and profound. Out of the very handicaps inherent in the medium he works in, he reconstructs a style which is also the life style of his people. His people dance, he dances. His people sing, he sings — in verse. His people stand with wearied heads, he brings them erect again. His people brave their broken bodies, he gives a swagger to them. And to the sport itself — to the most heavy-footed sport of all — he brings an artistry unsurpassed by Pavlova. Out of an elegy he makes a hymn.

Jonathan*

He was seven when they took his brother from him — his big brother — and put him in jail — because, or so they said, he had stolen $70 from a petrol station. The boy did not believe it: surely they would soon realise their mistake and send his brother back home to him.

But the years went by and George did not return. They kept him in prison on one pretext or another. Jonathan, now going on fourteen, became fearful for his brother. At the Catholic primary school to which his parents had sent him, hoping in their love to spare him the burden of his blackness, he had been a promising pupil and a popular one. At Pasadena High, Jonathan had fought his first battle: a boy had made foul remarks about black people. By the time he was transferred to Blair, the world had begun to close in on him: the grief of the ghetto had crept into his heart, his brother's predicament

* On Jonathan Jackson, in *Free the Soledad Brothers*, London, July 1971 (pamphlet).

possessed his mind. He saw that his parents and his sisters could not help: they were defeated at every turn, beaten back by the system, fobbed off by their congressmen. He looked for understanding to his teachers at Blair. They did not care. No one cared — and those who cared were powerless to do anything. Even Fay Stender, George's devoted lawyer, what could she do in the end? George was headed for the gas chamber.

In despair, Jonathan walked the courts of Los Angeles looking for a modicum of justice to assuage his disbelief. But he came back hurt, disillusioned, more wounded than before. 'That judge sits there, Momma, and says no to everything. He doesn't even stop to think before he says no.' And if there was no justice in the courts, where could a man turn to? George, Jonathan knew with a blinding certainty, was going to be killed: if the prisons did not get him, the courts would.

It all devolved on him now. He had to take up the burdens of justice and love and liberation — for himself, for George, for his people, for oppressed people everywhere. And so he stood there, that morning of 7 August 1970 — a boy barely turned seventeen — in the Marin County Courthouse, gun in hand, a judge as hostage (for how else could he set justice free?), demanding that the Soledad Brothers be freed by 12.30. Minutes later Jonathan was dead. But even as he died a moment of justice flickered like a miracle through America — and 'the moment of a miracle is unending lightning'.

Angela Davis*

Take a black child. Drop him in the ghetto. School him in the ways of poverty, discrimination and delinquency. When he graduates into crime and steals so much as $75, incarcerate him. If by some chance he rehabilitates himself in prison — and black rehabilitation means a total and complete understanding of the black condition — keep him there. Inveigle him into tawdry

* 'Angelus' in *Race Today*, vol. IV, no. 7, July 1972.

misdemeanours — keep him there. If out of his blackness he finds a larger political credo, stay him from parole, engineer misadventures for himself and his friends -- keep him there. If he then makes prison a massive *foco* of resistance, plan his death — keep him there no more, shoot him down.

And if somewhere in the world outside he has a friend, a lover, a brother, then bleed them *white*, railroad them into prison, harass them into the grave. If at the end of the year, or two, or more, the ordinary peoples of the world will not let you keep them in prison any longer, bring out the banners, fly high the American flag. Justice has been done.

Grotesque — the recipes for prison and the recipes for justice. Charades, macabre dances of death. And little punk white journalists caught up in the blind-bright light of their whiteness can't see the stark, dark world of the blacks. They have only to look, with whatever sensibilities they have left in them, at the hideous statistics of black oppression to see that the path to Angela's release is strewn with a host of the maimed and the murdered. They have only to look at Angela herself to see how in gathering and focusing in her own person the varied liberation movements of our time — as woman, as black, as *intellectuel engagé*, as communist — she alone could have roused the people of the world and her jurors to justice. She alone could have shown, with the abounding love of the true revolutionary, that power belongs to the people, that justice resides in their vigilance, that if they come for her in the morning...

Like George before her, out of her shackles she showed us chained, 'out of dust she made us into men'.

James Baldwin*

We try to forget Baldwin, or rather, we try to remember him as the author of the searing *Fire Next Time* or of that superbly

---

* Review of *James Baldwin: A Collection of Critical Essays*, edited by Kenneth Kinnamon, New Jersey, Prentice-Hall, 1974, in *Race & Class*, vol. XVII, no. 1, summer 1975.

documentary novel, *Go Tell it on the Mountain*. We even remember him marching with the King that God forgot, we remember him bear witness to the travail of his people, our people, their resistance and rebellion. We see him taking conscience of himself in his history and wanting to change it.

But when the fight for rights was over — for the vote and property and a place in the American sun — opening up the road ahead for the long and protracted and deathly struggle against capital and the state, against exploitation on a world scale, against death-dealing imperialism, Baldwin was not there. He was writing in Paris and London and Istanbul, or rapping pointlessly with Margaret Mead and her ilk, or making an arse of himself on BBC television — returning endlessly to the themes of love and redemption and the life interior: 'if we are going to change history... we have got to change ourselves because we are history'. True he rallied — momentarily — to the cause of Angela Davis, but without the faintest notion of what the larger issues were all about — an innocent abroad, from another country, another world, the world of the writer qua writer.

Baldwin himself had declared very early on in his career that he was a writer first and a black who happened to be American, after. But because what he wrote about — his condition and therefore that of his people — was of historical moment, he was caught up in the vortex of racial struggle. But as that phase yielded to class war Baldwin remained stranded on the shores of another time. His writings became insubstantial, devoid of content, of historical moment. All that was left was his craft.

Kinnamon's collection of essays is designed to provide a literary assessment of Baldwin. They have been culled mostly from American journals ranging over a period of time — from Langston Hughes in 1956 to De Mott in 1972. They examine Baldwin's skills as a writer rather more than the substance of his writings, or the substance only in terms of the form. 'To function as a voice of outrage month after month for a decade or more strains the heart and mind, and rhetoric as well; the consequence is a writing style ever on the edge of being winded by too many summonses to intensity' (De Mott). Or 'his colour is his metaphor' (Newman).

In fact, though, his colour is his stock-in-trade, Baldwin's obsession is with himself, and with America and race only when that country's history is being acted out in him. And when that history is in racial crisis, no one portrays its paroxysms better than Baldwin — and that by writing about himself, his pain, his despair, his quest for love and acceptance. At that point the objective condition and the subjective become one, identical, and Baldwin bears witness to that history as no one can. What Mailer attempts Baldwin is.

Kinnamon's own assessment of Baldwin as 'victim, witness and prophet' is therefore a truer judgement. But only Cleaver in his 'Notes on a Native Son' makes a social, even political, assessment of Baldwin — and inevitably he summons up the revolutionary spectre of Richard Wright, whom Baldwin bitched about in *Nobody knows my name*, to beat Baldwin with. Nearly all the other contributors attempt to locate Baldwin in the bourgeois literary constellation of Henry James, Hart Crane and Hemingway.

The point, however, is not whether Baldwin is a consummate craftsman or not — that he is — or whether he has words or not — that he has — but whether he is a writer who happens to be black or a black man who happens to be a writer. The one seeks to liberate himself through his people, the other to liberate his people through him. The one is a professional, a mercenary on hire to his people, the other is a soldier in the people's army.

The Colony of the Colonised: Notes on Race, Class and Sex*

None of the movements triggered off by the black liberation struggles of the last decade or two offers so sharply to clarify the issues of the black struggle itself as the women's liberation

* *Race Today*, vol. V, no. 6, June 1973.

movement. The Chicano, Puerto Rican and Amerindian movements were, in effect, variations on the theme of nationalist struggle, of struggle as a people — because it was as a people (and not as a class) that they had perceived their oppression in the first instance. But this served only to replicate the problems of analysis, strategy and organisation of the black struggle without in any way helping to clarify them. In the event, these movements ended up by being integrated into capitalism's plural society or, on those occasions when they wandered off into the class struggle, by being subsumed to white working-class strategy. In either case the opportunity for a vanguard role in revolutionary politics had been lost to them.

Part of the reason for this surrender, particularly on the part of the black movement, can be traced to the confused historical position in which black people find themselves placed in white capitalist society. Whereas the working class sees itself exploited as a class and comes face to face with its exploiter, capital, the capitalist exploitation of blacks is veiled by racial oppression. As a result, they are caught up in a two-fold consciousness: as a class and as a race, each of which often contradicts the other without affording a synthesis.

But this again is due to the fact that, in their thinking black theoreticians have been unable to make a concrete analysis of their concrete situation without using the categories of Eurocentric marxism. They have, that is, been unable to apply the tools of marxist analysis to their own specific situation without borrowing at the same time concepts relevant to another time and another experience. Hence they have alternated between class theories and caste theories without being able to fuse the two. Only in the concept of black revolutionary nationalism of the Black Panther Party has emerged even the basis of such a synthesis — and in the abortive attempts of the Black Workers' Revolutionary Movement to lead white workers.

The women's liberation movement, on the other hand, by the very nature of its struggle, has been forced to question every man-made theory and tactic, marxist and otherwise. Faced with an oppression which she is led to believe is functional to her sex and with an exploitation which she is led to believe is only contingent on the exploitation of her man, the woman has had to seek her liberation from both caste and class simultaneously — and in their totality. Released from her class, she may still be oppressed as a woman, released from her caste she may still be exploited in her class. Even when she is not exploited as a wage slave in a factory, she is exploited in her home as the slave of a wage-slave; even when she is not producing commodity for capital, she is producing the labourer who will produce the commodity that capital requires. Or she is being exploited in both places at once, producing and reproducing for capital at once. In her caste and in her class, in the work place and at home, her role in the cycle of social production is unending and total.

Thus her exploitation and her oppression are both direct and derivative, overt and concealed, but in bringing them to the surface and in confronting them directly she is forced to address herself to the specificity of her problem without help from anyone.

Unlike the blacks, women cannot be said to have even a life-style of their own. Everything the woman has got by way of culture has been handed down to her by her oppressor and must necessarily implicate her in her oppression. Even the concepts she may use for an understanding of her position are, in the final analysis, particular to men. And the tools she seeks for her liberation are not particular to her struggle. The interpretations, analyses, strategies of other liberation movements avail her very little. The struggles of the blacks are the struggles of a colonised people; she is a colony of the colonised. They may afford her parallels or take her part of the way, but never any more than that.

And so she is thrown back on her own experience, her own perceptions and observations — and from these she must derive the concepts and theories that she must apply to her own particular form of struggle. But in a way this is an advantage. For most liberation movements in the west have tended at one time or another to misplace themselves in the revolutionary experience of other peoples and to borrow theories and practices which had little relevance to the conditions of their own struggle. Even the black American struggle failed to fulfil its potential role as a revolutionary vanguard precisely because it did not use the specificity of its experience — as a nation and as a class both at once — to redefine class and the class struggle itself.

On the other hand, the women's movement, because it begins with a re-definition of woman and is constantly on guard against being defined out of itself again, is forced to keep the specificity — and the autonomy — of its struggle constantly in mind. The counter-revolution is not something that merely infiltrates her politics, but a very real existential danger with its counterpart in politics. Caught up in a world that men have made, she must continuously construct and re-construct her reality — in herself and in her struggle, subjectively and objectively — in a daily act of renewal, as it were, and in renewed struggle. It is a sort of permanent revolution in which merely to hold on to the reality which the woman finds for herself constitutes a revolutionary act — if only because that really is a refusal to accept the world as defined for her, a refusal to accept the status quo.

It may be that not all things that men have made are bad; it may be that the revolutionary struggles that men have entered into have something to tell her about her own; it may even be that in the final analysis the only way to expect change in society is through class struggle. But whatever the truth of the matter, the woman must come to it herself, relate it to the specific experience of her own specific oppression and exploitation — travel from the particular experience to the general, from caste to class, and in the process bring to the class struggle itself some of the perspectives and directions and leadership which the black struggle had finally failed to give it and which it so sorely requires.

Western working class movements have belied their own class experience by refusing to re-assess both their class instinct and their class position in the light of new historical forces, especially the forces of revolution removed from their own. Partly this has been due to the rigours of working class dogma; but it has also been due to the working class acceptance of capitalist cultural norms. And some aspects of this culture such as racial superiority and male supremacy are so ingrained in the white working class that any attack on them is either explained away as an aberration from the norm or is reduced to its basic economic causes and subsumed to the principal economic struggle of class. Thus denied its own dynamic of struggle, the cultural rebellion is prevented from bringing to the class struggle valid new insights and perspectives. It, in turn, loses its political thrust. As a result capitalism is able to absorb and negate the cultural revolt while at the same time perpetuating its cultural hegemony over the working class.

To put it differently, in concentrating so exclusively on the economic determinants of the class struggle, the working class has ignored the cultural superstructure which not only reinforces the capitalist base but produces its own categories of oppression and exploitation which further divide the working class. The culture of racial superiority puts the white worker in a 'class' above the woman. In both cases the exploited are used to further the exploitation of blacks and of women and thereby reinforce the capitalist economy which in turn exploits the worker.

Quite clearly the economic and cultural aspects of struggle — the standard of living and the quality of life — are interdependent and retroactive. But capitalism over the years has been able not only to divide and separate the two but to conceal them from each other. In the process one aspect of the struggle passed on to the working class, and the other to sections of society denoted as caste — and the totality of the struggle was lost to view.

The black movement, as has already been observed, tried to grasp this totality and failed because it was unable to redefine the class struggle in the light of its own history. All too conscious of his position as a black wage-slave within capitalism, the black man has lost sight of that unique period of his history in which he had been designated a thing, a commodity and therefore outside the scheme of the social relations of production. In his preoccupation with the present he had lost out on his past or misplaced it in Africa. He was denied, for the nonce, that historical sense which 'involves a perception not only of the pastness of the past, but of the present'.[1]

Women in the home, on the other hand, have been considered as use-values throughout history — not as producers of surplus value. Left out of the class-reckoning, it was inevitable that in defining themselves women would redefine the nature of class itself. Already such a redefinition has begun to emerge in a brilliant pamphlet by Mariarosa Dalla Costa and Selma James who 'pose as foremost the need to break this role of housewife that wants woman divided from each other, from men, and from children, each locked in her family as the chrysalis in the cocoon that imprisons itself by its own work, to die and leave silk for capital'.[2]

Power to the sisters and therefore to the class.

Notes

1. T.S. Eliot, 'Tradition and the Individual Talent', in *The Sacred Wood: Essays on Poetry and Criticism*, London: Methuen, 1934.
2. Mariarosa Dalla Costa and Selma James, *The Power of Women and the Subversion of the Community*, Bristol: Falling Wall Press, 1973.

Paul Robeson*

In a remote village in the north of Ceylon many years ago, a group of boys, playing truant from school, crowded into the village bakery to look at their first wireless. The owner twiddled the knobs with a flourish, showing his audience how he could bring the world to his doorstep. And suddenly he stopped — at an English song — though he understood not a word. A man was singing what sounded like a song of his people that sounded so much like their own — and he sang as though the big heart of the radio itself would break. And they all fell silent, as though in prayer.

I was one of those boys. The man's name escaped me then, but I was to stumble across it some years later in an essay by Alexander Woolcott called 'Colossal Bronze'. And I knew it was him, not just because Woolcott was describing a great singer ('the finest musical instrument wrought by nature in our time'), but because he was trying to connect the singing to some other quality in the man from which that singing sprang, and which we had all fallen silent to that day in the bakery — a sort of universality that connected us to him and was 'coeval with Adam and the redwood trees of California'. 'By what he does, thinks and

---

* Review of *Paul Robeson Speaks: Writings, Speeches, Interview, 1918-74*, edited by Philip S. Foner, London: Quartet 1978, in *Race & Class*, vol. XXI, no. 3, Winter 1980.

is,' wrote Woolcott, 'by his unassailable dignity and his serene incorruptible simplicity, Paul Robeson strikes me as having been made out of the original stuff of the world... He is a fresh act, a fresh gesture, a fresh effort of creation.' It was a description that might have been considered extravagant had it not been measured out to the measure of the man himself.

But not since then — and Woolcott was writing in the 1930s when Robeson's life was barely half-way old, his endeavours but half-fulfilled — has anyone captured the size of the man as Philip Foner has done in this monumental work. And he does it not with fabulous language or metaphor, but with a meticulously researched, carefully documented, clinical presentation of Robeson through his own speeches, writings and interviews. In fact, Foner does not present Robeson so much as allow Robeson to present himself. Even in the introductory essay, where editors are wont to fly the flag of their disposition, Foner does not intrude his opinions. Instead, he gives us findings, from the evidence in his book. It is indeed Robeson who speaks, no other, and no one can henceforth say that there is another Robeson. Robeson stands defined. This is the definitive Robeson — not least because, if it is Robeson who speaks, it is Foner who, with scholarly precision and infinite care, provides the notes and references that place Robeson firmly in the politics and history of our time.

Long before black was officially beautiful, Robeson was celebrating the pride of his race and the cultures of his peoples. Long before the Civil Rights movement had taken root or the Black Power movement begun, Robeson was leading the battle against black second-class citizenship, challenging repressive laws, protesting the injustices of the courts, demonstrating on behalf of black activists and calling for boycotts, pickets and mass mobilisation. And civil rights did not mean just black civil rights, but civil rights *per se*. So he took on the McCarthyite HUAC and virtually asked them to get stuffed.

Even before the era of de-colonisation had taken off, Robeson was demanding freedom for the colonies. As co-founder and chairman of the Council of African Affairs (1937-55) he agitated for African liberation, mounted a vitriolic campaign against Malan and South Africa's 'foul creed' and revived interest in African art and culture. (It is little wonder then that Nkrumah should have offered him the chair in drama and music in the university of free Ghana.)

In the early 1930s Robeson was inveighing against anti-semitism and Hitler. In 1938 he was in Spain denouncing fascism. In the 1940s he condemned American arms supplies to the Dutch in Indonesia and to the French in Indo-China. In 1950 he came out against the Korean war. In 1954, long before the anti-war movement, he likened Ho Chi Minh to Toussaint L'Ouverture as the liberator of his people and demanded that the 'imperialists be stopped in their tracks'. At the height of the cold war, he urged detente and peace. In the heart of monopoly capital he preached socialism.

But nowhere in the annals of American history, black or white, up to the writing of this book, has Robeson been acknowledged as a political activist. Even the black political movements of the 1960s, though acknowledging their debt to Martin Luther King and Malcolm X, have failed to see Robeson as the forerunner of them all. (That he was still alive at the time, though in comparative retirement, makes that neglect even sadder.)

Most writers have been content to write of Robeson as a singer or an actor. (He was also a Phi Beta Kappa scholar, a lawyer and an All-American footballer.) But song, for Robeson, and in particular folk song, was what connected him to his own sharecropper origins and the ordinary peoples of the world — and committed him to their struggles. The song for him was the singer, the artist his art — but singer or artist, he was nothing without the people. For 'when as a singer I walk on to the platform, to sing back to the people the songs they themselves have created, I can feel a great unity, not only as a person, but as an artist who is one with his audience'. And it is that unity, that wholeness, that Robeson brought to his politics too, so that even song became merely an instrument in the battle, art his weapon. They would ban him from the concert hall and stage, but he could still sing his politics; they would take away his passport, but he could still reach out to audiences across the world — over the telephone. They could not imprison his voice, and when they did, betwixt times, he wrote — column after militant column in his *Freedom* newspaper. But that did not mean he considered himself a writer, for that too was but an instrument in the struggle.

The purpose of life for Robeson was to be free. But he himself could not be free till all men were free, and that, in concrete terms, meant the oppressed and the exploited. So he put his own freedom on the line — for them and, therefore, himself. He was as basic as that, and as universal — 'a man and a half' as Ossie Davis put it.

The Liberation of the Black Intellectual*

> Wherever colonisation is a fact, the indigenous culture begins to rot. And among the ruins something begins to be born which is not a culture but a kind of sub-culture, a subculture which is condemned to exist on the margin allowed by an European culture. This then becomes the province of a few men, the elite, who find themselves placed in the most artificial conditions, deprived of any revivifying contact with the masses of the people.
>
> *Aimé Césaire*[1]

On the margin of European culture, and alienated from his own, the 'coloured' intellectual is an artefact of colonial history, marginal man par excellence. He is a creature of two worlds, and of none. Thrown up by a specific history, he remains stranded on its shores even as it recedes. And what he comes into is not so much a twilight world, as a world of false shadows and false light.

At the height of colonial rule, he is the servitor of those in power, offering up his people in return for crumbs of privilege; at its end, he turns servant of the people, negotiating their independence even as he attains to power. Outwardly, he favours that part of him which is turned towards his native land. He puts on the garb of nationalism, vows a return to tradition. He

---

* First published as 'Alien Gods' in *Colour Culture and Consciousness*, London: Allen & Unwin, 1974.

helps design a national flag, compose a People's Anthem. He puts up with the beat of the tom-tom and the ritual of the circumcision ceremony. But privately, he lives in the manner of his masters, affecting their style and their values, assuming their privileges and status. And for a while he succeeds in holding these two worlds together, the outer and the inner, deriving the best of both. But the forces of nationalism on the one hand and the virus of colonial privilege on the other, drive him once more into the margin of existence. In despair he turns himself to Europe. With something like belonging, he looks towards the Cathedral at Chartres and Windsor Castle, Giambologna and Donizetti and Shakespeare and Verlaine, snow-drops and roses. He must be done, once and for all, with the waywardness and uncouth manners of his people, released from their endemic ignorance, delivered from witchcraft and voodoo, from the heat and the chattering mynah-bird, from the incessant beat of the tom-tom. He must return to the country of his mind.

But even as the 'coloured' intellectual enters the mother country, he is entered into another world where his colour, and not his intellect or his status, begins to define his life — he is entered into another relationship with himself. The porter (unless he is black), the immigration officer (who is never anything but white), the customs official, the policeman of whom he seeks directions, the cabman who takes him to his lodgings, and the landlady who takes him in at a price — none of them leaves him in any doubt that he is not merely not welcome in their country, but should in fact be going back — to where he came from. That indeed is their only curiosity, their only interest: where he comes from, which particular jungle, Asian, African or Caribbean.

There was a time when he had been received warmly, but he was at Oxford then and his country was still a colony. Perhaps equality was something that the British honoured in the abstract. Or perhaps his 'equality' was something that was precisely defined and set within the enclave of Empire. He had a place somewhere in the imperial class structure. But within British society itself there seemed no place for him. Not even his upper-class affectations, his BBC accent, his well-pressed suit and college tie afford him a niche in the carefully defined inequalities of British life.

He feels himself not just an outsider or different, but invested, as it were, with a separate inequality: outside and inferior at the same time.

At that point, his self-assurance which had sat on him 'like a silk hat on a Bradford millionaire'[2] takes a cruel blow. But he still has his intellect, his expertise, his qualifications to fall back on. He redeems his self-respect with another look at his Oxford diploma (to achieve which he had put his culture in pawn). But his applications for employment remain unanswered, his letters of introduction unattended. It only needs the employment officer's rejection of his qualifications, white though they be, to dispel at last his intellectual pretensions.

The certainty finally dawns on him that his colour is the only measure of his worth, the sole criterion of his being. Whatever his claims to white culture and white values, whatever his adherence to white norms, he is first and last a no-good nigger, a bleeding wog or just plain black bastard. His colour is the only reality allowed him; but a reality which, to survive, he must cope with. Once more he is caught between two worlds: accepting his colour and rejecting it, or accepting it only to reject it — aping still the white man (though now with conscious effort at survival), playing the white man's game (though now aware that he changes the rules so as to keep on winning), even forcing the white man to concede a victory or two (out of his hideous patronage, his grotesque paternalism). He accepts that it is their country and not his, rationalises their grievances against him, acknowledges the chip on his shoulder (which he knows is really a beam in their eye), and, ironically, by virtue of staying in his place, moves up a position or two — in the area, invariably, of race relations.[3] For it is here that his skilled ambivalence finds the greatest scope, his colour the greatest demand. Once more he comes into his own — as servitor of those in power, a buffer between them and his people, a shock-absorber of 'coloured discontent' — in fact, a 'coloured' intellectual.

But this is an untenable position. As the racial 'scene' gets worse, and racism comes to reside in the very institutions of white society, the contradictions inherent in the marginal situation of the 'coloured' intellectual begin to manifest themselves. As a 'coloured' he is outside white society, in his intellectual functions he is outside black. For if, as Sartre has pointed out, 'that which defines an intellectual... is the profound contradiction between the universality which bourgeois society is obliged to allow his scholarship, and the restricted ideological and political domain in which he is forced to apply it',[4] there is for the 'coloured' intellectual no role in an 'ideological and political domain', shot through with racism, which is not fundamentally antipathetic to his colour and all that it implies. But for that very reason, his contradiction, in contrast to that of his white counterpart, is perceived not just intellectually or abstractly, but in his very existence. It is for him, a living, palpitating reality, demanding resolution.

Equally, the universality allowed his scholarship is, in the divided world of a racist society, different to that of the white intellectual. It is a less universal universality, as it were, and subsumed to the universality of white scholarship. But it is precisely because it is a universality that is particular to colour that it is already keened to the sense of oppression. So that when Sartre tells us that the intellectual, in grasping his contradictions, puts himself on the side of the oppressed ('because, in principle, universality is on that side'),[5] it is clear that the 'coloured' intellectual, at the moment of grasping his contradictions, *becomes* the oppressed — is reconciled to himself and his people, or rather, to himself in his people.

To put it differently. Although the intellectual qua intellectual can, in 'grasping his contradiction', take the *position* of the oppressed, he cannot, by virtue of his class (invariably petty-bourgeois) achieve an instinctual understanding of oppression. The 'coloured' man, on the hand, has, by virtue of his colour, an *instinct* of oppression, unaffected by his class, though muted by it. So that the 'coloured' intellectual, in resolving his contradiction as an intellectual, resolves also his existential contradiction. In coming to consciousness of the oppressed, he 'takes conscience of himself',[6] in taking conscience of himself, he comes to consciousness of the oppressed. The fact of his intellect which had alienated him from his people now puts him on their side, the fact of his colour which had connected him with his people, restores him finally to their ranks. And at that moment of reconciliation between instinct and position, between the existential and the intellectual, between the subjective and objective realities of his oppression, he is delivered from his marginality and stands revealed as neither 'coloured' nor 'intellectual' — but BLACK.[7]

He accepts now the full burden of his colour. With Césaire, he cries:

> I accept... I accept... entirely, without
> reservation...
> my race which no ablution of hyssop mingled with
> lilies can ever purify
>
> my race gnawed by blemishes
> my race ripe grapes from drunken feet
> my queen of spit and leprosies
> my queen of whips and scrofulae
> my queen of squamae and chlosmae
> (O royalty whom I have loved in the far gardens of
> spring lit by chestnut candles!)
> I accept. I accept.[8]

And accepting, he seeks to define. But black, he discovers, finds definition not in its own right but as the opposite of white.[9] Hence in order to define himself, he must first define the white man. But to do so on the white man's terms would lead him back to self-denigration. And yet the only tools of intellection available to him are white tools — white language, white education, white systems of thought — the very things that alienate him from himself. Whatever tools are native to him lie beyond his consciousness somewhere, condemned to disuse and decay by white centuries. But to use white tools to uncover the white man so that he (the black) may at last find definition requires that the tools themselves are altered in their use. In the process, the whole of white civilisation comes into question, black culture is re-assessed, and the very fabric of bourgeois society threatened.

Take language, for instance. A man's whole world, as Fanon points out, is 'expressed and implied by his language':[10] it is a way of thinking, of feeling, of be-ing. It is identity. It is, in Valéry's grand phrase, 'the god gone astray in the flesh'.[11] But the language of the colonised man is another man's language. In fact it is his oppressor's and must, of its very nature, be inimical to him — to his people and his gods. Worse, it creates alien gods. Alien gods 'gone astray in the flesh' — white gods in black flesh — a canker in the rose. No, that is not quite right, for white gods, like roses, are beautiful things, it is the black that is cancerous. So one should say a 'rose in the canker'. But that is not quite right either — neither in its imagery nor in what it is intended to express. How does one say it then? How does one express the holiness of the heart's disaffection (*pace* Keats) and 'the truth of the imagination' in a language that is false to one? How does one communicate the burden of one's humanity in a language that dehumanises one in the very act of communication?

Two languages, then, one for the coloniser and another for the colonised — and yet within the same language? How to reconcile this ambivalence? A patois, perhaps: a spontaneous, organic rendering of the masters' language to the throb of native sensibilities — some last grasp at identity, at wholeness.

But dialect betrays class. The 'pidgin-nigger-talker' is an ignorant man. Only common people speak pidgin. Conversely, when the white man speaks it, it is only to show the native how common he really is. It is a way of 'classifying him, imprisoning him, primitivising him'.[12]

Or perhaps the native has a language of his own; even a literature. But compared to English (or French) his language is dead, his literature passé. They have no place in a modern, industrialised world. They are for yesterday's people. Progress is English, education is English, the good things in life (in the world the coloniser made) are English, the way to the top (and white civilisation leaves the native in no doubt that that is the purpose of life) is English. His teachers see to it that he speaks it in school, his parents that he speaks it at home — even though they are rejected by their children for their own ignorance of *the* tongue.

But if the coloniser's language creates an 'existential deviation'[13] in the native, white literature drives him further from himself. It disorientates him from his surroundings: the heat, the vegetation, the rhythm of the world around him. Already, in childhood, he writes school essays on 'the season of mists and mellow fruitfulness'. He learns of good and just government from Rhodes and Hastings and Morgan. In the works of the great historian, Thomas Carlyle, he finds that 'poor black Quashee... a swift supple fellow, a merry-hearted, grinning, dancing, singing, affectionate kind of creature' could indeed be made into a 'handsome glossy thing' with a 'pennyworth of oil', but 'the tacit prayer he makes (unconsciously he, poor blockhead) to you and to me and to all the world who are wiser than himself is "compel me" ' — to work.[14] In the writings of the greatest playwright in the world, he discovers that he is Caliban and Othello and Aaron, in the testaments of the most civilised religion that he is for ever cursed to slavery. With William Blake, the great revolutionary poet and painter, mystic and savant, he is convinced that:

> My mother bore me in the southern wild,
> And I am black, but O! my soul is white;
> White as an angel is the English child,
> But I am black, as if bereav'd of light.[15]

Yet, this is the man who wrote 'The Tyger'. And the little black boy, who knows all about tigers, understands the great truth of Blake's poem, is lost in wonderment at the man's profound imagination. What then of the other Blake? Was it only animals he could imagine himself into? Did he who wrote 'The Tyger' write 'The Little Black Boy'?

It is not just the literature of the language, however, that ensnares the native into 'whititude', but its grammar, its syntax, its vocabulary. They are all part of the trap. Only by destroying the trap can he escape it. 'He has', as Genet puts it, 'only one recourse: to accept this language but to corrupt it so skilfully that the white men are caught in his trap.'[16] He must blacken the language, suffuse it with his own darkness, and liberate it from the presence of the oppressor.

In the process, he changes radically the use of words, word-order — sounds, rhythm, imagery — even grammar. For, he recognises with Laing that even 'syntax and vocabulary are political acts that define and circumscribe the manner in which facts are experienced, [and] indeed... create the facts that are studied'.[17] In effect he brings to the language the authority of his particular experience and alters thereby the experience of the language itself. He frees it of its racial oppressiveness (black *is* beautiful), invests it with 'the universality inherent in the human condition'.[18] And he writes:

> As there are hyena-men and panther-men
> so I shall be a Jew man
> a Kaffir man
> a Hindu-from-Calcutta man
> a man-from-Harlem-who-hasn't-got-the-vote.[19]

The discovery of black identity had equated the 'coloured' intellectual with himself, the definition of it equates him with all men. But it is still a definition arrived at by negating a negative, by rejecting what is not. And however positive that rejection, it does not by itself make for a positive identity. For that reason, it tends to be self-conscious and overblown. It equates the black man to other men on an existential (and intellectual) level, rather than on a political one.

But to 'positivise' his identity, the black man must go back and rediscover himself — in Africa and Asia — not in a frenetic search for lost roots, but in an attempt to discover living tradition and values. He must find, that is, a historical sense, 'which is a sense of the timeless as well as the temporal, and of the timeless and temporal together'[20] and which 'involves a perception not only of the pastness of the past, but of its presence'.[21] Some of that past he still carries within him, no matter that it has been mislaid in the Caribbean for over four centuries. It is the presence of that past, the living presence, that he now seeks to discover. And in discovering where he came from he realises more fully where he is at, and where, in fact, he is going to.

He discovers, for instance, that in Africa and Asia, there still remains, despite centuries of white rule, an attitude towards learning which is simply a matter of curiosity, a quest for understanding — an understanding of not just the 'metalled ways' on which the world moves, but of oneself, one's people, others whose life styles are alien to one's own — an understanding of both the inscape and fabric of life. Knowledge is not a goal in itself, but a path to wisdom; it bestows not privilege so much as duty, not power so much as responsibility. And it brings with it a desire to learn even as one teaches, to teach even as one learns. It is used not to compete with one's fellow beings for some unending standard of life, but to achieve for them, as for oneself, a higher quality of life.

'We excel', declares the African,

> neither in mysticism nor in science and technology, but in the field of human relations... By loving our parents, our brothers, our sisters and cousins, aunts, uncles, nephews and nieces, and by regarding them as members of our families, we cultivate the habit of loving lavishly, of exuding human warmth, of compassion, and of giving and helping... Once so conditioned, one behaves in this way not only to one's family, but also to the clan, the tribe, the nation, and to humanity as a whole.[22]

Chisiza is here speaking of the unconfined nature of love in African (and Asian) societies (not, as a thousand sociologists would have us believe, of 'the extended family system'), in marked contrast to western societies where the love between a man and a woman (and their children) is sufficient unto itself, seldom opening them out, albeit through each other, to a multitude of loves. The heart needs the practice of love as much as the mind its thought.

The practical expression of these values is no better illustrated than in the socialist policies of Nyerere's Tanzania. It is a socialism particular to African conditions, based on African tradition, requiring an African (Swahili) word to define it. *Ujamaa* literally means 'familyhood'. 'It brings to the mind of our people the idea of mutual involvement in the family as we know it.'[23] And this idea of the family is the sustaining principle of Tanzanian society. It stresses cooperative endeavour rather than individual advancement. It requires respect for the traditional knowledge and wisdom of one's elders, illiterate though they be, no less than for academic learning. But the business of the educated is not to fly away from the rest of society on the wings of their skills, but to turn those skills to the service of their people. And the higher their qualifications, the greater their duty to serve. 'Intellectual arrogance', the Mwalimu has declared, 'has no place in a society of equal citizens.'[24]

The intellectual, that is, has no special privilege in such a society. He is as much an organic part of the nation as anyone else. His scholarship makes him no more than other people and his functions serve no interest but theirs. There is no dichotomy here between status and function. Hence he is not presented with the conflict between the universal and the particular of which Sartre speaks. And in that sense he is not an intellectual but everyman.

The same values obtain in the societies of Asia, sustained not so much by the governments of the day as in the folklore and tradition of their peoples. The same sense of 'family-hood', of the need to be confirmed by one's fellow man, the notion of duty as opposed to privilege, the preoccupation with truth rather than fact and a concept of life directed to the achievement of unity in diversity, characterise the Indian ethos. One has only to look at Gandhi's revolution to see how in incorporating, in its theory and its practice, the traditions of his people, a 'half-naked fakir' was able to forge a weapon that took on the whole might of the British empire and beat it. Or one turns to the early literature and art of India and finds there that the poet is less important than his poem, the artist more anonymous than his art. As Benjamin Rowland remarks: 'Indian art is more the history of a society and its needs than the history of individual artists.'[25] The artist, like any other individual, intellectual or otherwise, belongs to the community, not the community to him. And what he conveys is not so much his personal experience of truth as the collective vision of a society of which he is part, expressed not in terms private to him and his peers, but in familiar language — or in symbols, the common language of truth.

In western society, on the other hand, art creates its own coterie. It is the province of the specially initiated, carrying with it a language and a life-style of its own, even creating its own society. It sets up cohorts of interpreters and counter-interpreters, middlemen, known to the trade as *critics*, who in disembowelling his art show themselves more powerful, more creative than the artist. It is they who tell the mass of the people how they should experience art. And the more rarified it is, more removed from the experience of the common people, the greater is the artist's claim to *art* and the critic's claim to authority. Did but the artist speak directly to the people and from them, the critic would become irrelevant, and the artist symbiotic with his society.

It is not merely in the field of art, however, that western society shows itself fragmented, inorganic and expert-oriented. But the fact that it does so in the noblest of man's activities is an indication of the alienation that such a society engenders in all areas of life. In contrast to the traditions of Afro-Asian countries, European civilisation appears to be destructive of human love and cynical of human life. And nowhere do these traits manifest themselves more clearly than in the attitude towards children and the treatment of the old. Children are not viewed as a challenge to one's growth, the measure of one's possibilities, but as a people apart, another generation, with other values, other standards, other aspirations. At best one keeps pace with them, puts on the habit of youth, feigns interest in their interests, but seldom if ever comprehends them. Lacking openness and generosity of spirit, the ability to live dangerously with each other, the relationship between child and adult is rarely an organic one. The adult occupies the world of the child far more than the child occupies the world of the adult. In the result, the fancy and innocence of children are crabbed and soured by adulthood even before they are ready to beget choice.

Is it any wonder then that this tradition of indifference should pass on back to the old from their children? But it is a tradition that is endemic to a society given to ceaseles competition and ruthless rivalry — where even education is impregnated with the violence of divisiveness, and violence itself stems not from passion (an aspect of the personal), but from cold and calculated reason (an aspect of the impersonal). When to get and to spend is more virtuous than to be and to become, even lovers cannot abandon themselves to each other, but must work out the debit and credit of emotion, a veritable balance-sheet of love. Distrust and selfishness and hypocrisy in personal relationships, and plain cruelty and self-aggrandisement in the art of government are the practice of such a society, however elevating its principles. Government itself is the art of keeping power from the people under the guise of the people's will. And the working people themselves are inveigled into acquiescence of the power structure by another set of middlemen: the union bosses.

In the face of all this, the black man in a white society — the black man, that is, who has 'taken conscience of himself', established at last a positive identity — comes to see the need for radical change in both the values and structure of that society. But even the revolutionary ideologies that envisage such a change are unable to take into their perspective the nature of his particular oppression and its implications for revolutionary strategy. White radicals continue to maintain that colour oppression is no more than an aspect of class oppression, that colour discrimination is only another aspect of working-class exploitation, that the capitalist system is the common enemy of the white worker and black alike. Hence they require that the colour line be subsumed to the class line and are satisfied that the strategies worked out for the white proletariat serve equally the interests of the black. The black strugle, therefore, should merge with and find direction from the larger struggle of the working class as a whole. Without white numbers, anyway, the black struggle on its own would be unavailing.

But what these radicals fail to realise is that the black man, by virtue of his particular oppression, is closer to his bourgeois brother (by colour) than to his white comrade. Indeed his white comrade is a party to his oppression. He too benefits from the exploitation of the black man, however indirectly, and tends to hold the black worker to areas of work which he himself does not wish to do, and from areas of work to which he himself aspires, irrespective of skill. In effect, the black workers constitute that section of the working class which is at the very bottom of society and is distinguished by its colour. Conversely, the attitude of racial superiority on the part of white workers relegates their black comrades to the bottom of society. In the event, they come to constitute a class apart, an under-class: the sub-proletariat. And the common denominator of capitalist oppression is not sufficient to bind them together in a common purpose.

A common understanding of racial oppression, on the other hand, ranges the black worker on the side of the black bourgeois against their common enemy: the white man, worker and bourgeois alike.

In terms of analysis, what the white marxists fail to grasp is that the slave and colonial exploitation of the black peoples of the world was so total and devastating — and so systematic in its devastation — as to make mock of working-class exploitation. Admittedly, the economic aspects of colonial exploitation may find analogy in white working-class history. But the cultural and psychological dimensions of black oppression are quite unparalleled. For, in their attempt to rationalise and justify to their other conscience 'the robbery, enslavement, and continued exploitation of their coloured victims all over the globe',[26] the conquistadors of Europe set up such a mighty edifice of racial and cultural superiority, replete with its own theology of goodness, that the natives were utterly disoriented and dehumanised. Torn from their past, reified in the present, caught for ever in the prison of their skins, accepting the white man's definition of themselves as 'quintessence of evil... representing not only the absence of values but the negation of values... the corrosive element disfiguring all that has to do with beauty or morality',[27] violated and sundered in every aspect of their being, it is a wonder that, like lemmings, they did not throw themselves in the sea. If the white workers' lot at the hands of capitalism was alienation, the blacks underwent complete deracination. And it is this factor which makes black oppression qualitatively different from the oppression of the white working class.

The inability of white marxists to accept the full import of such an analysis on the part of black people may be alleged to the continuing paternalism of a culture of which they themselves are victims. (Marxism, after all, was formulated in an European context and must, on its own showing, be Eurocentric.) Or it may be that to understand fully the burden of blackness, they require the imagination and feeling systematically denied them by their culture. But more to the point is that, in their preoccupation with the economic factors of capitalist oppression, they have ignored the importance of its existential consequences, in effect its consequences to culture. The whole structure of white racism is built no doubt on economic exploitation, but it is cemented with white culture. In other words, the racism inherent in white society is *determined* economically, but *defined* culturally. And any revolutionary ideology that is relevant to the times must envisage not merely a change in the ownership of the means of production, but a definition of that ownership: who shall own, whites only or blacks as well? It must envisage, that is, a fundamental change in the concepts of man and society contained in white culture — it must envisage a revolutionary culture. For, as Gramsci has said, revolutionary theory requires a revolutionary culture.

But to revolutionise a culture, one needs first to make a radical assessment of it. That assessment, that revolutionary perspective, by virtue of his historical situation, is provided by the black man. For it is with the cultural manifestations of racism in his daily life that he must contend. Racial prejudice and discrimination, he recognises, are not a matter of individual attitudes, but the sickness of a whole society carried in its culture. And his survival as a *black* man in white society requires that he constantly questions and challenges every aspect of white life even as he meets it. White speech, white schooling, white law, white work, white religion, white love, even white lies — they are all measured on the touchstone of his experience. He discovers, for instance, that white schools make for white superiority, that white law equals one law for the white and another for the black, that white work relegates him to the worst jobs irrespective of skill, that even white Jesus and white Marx who are supposed to save him are really not in the same street, so to speak, as black Gandhi and black Cabral. In his everyday life he fights the particulars of white cultural superiority, in his conceptual life he fights the ideology of white cultural hegemony. In the process he engenders not perhaps a revolutionary culture, but certainly a revolutionary practice within that culture.

For that practice to blossom into a revolutionary culture, however, requires the participation of the masses, not just the blacks. This does not mean, though, that any ad hoc coalition of forces would do. Coalitions, in fact, are what will not do. Integration, by any other name, has always spelt death — for the blacks. To integrate with the white masses before they have entered into the practice of cultural change would be to emasculate the black cultural revolution. Any integration at this stage would be a merging of the weaker into the stronger, the lesser into the greater. The weakness of the blacks stems from the smallness of their numbers, the 'less-ness' from the bourgeois cultural consciousness of the white working class.

Before an organic fusion of forces can take place, two requirements need to be fulfilled. The blacks must through the consciousness of their colour, through the consciousness, that is, of that in which they perceive their oppression, arrive at a consciousness of class; and the white working class must in recovering its class instinct, its sense of oppression, both from technological alienation and a white-oriented culture, arrive at a consciousness of racial oppression.

For the black man, however, the consciousness of class is instinctive to his consciousness of colour.[28] Even as he begins to throw away the shackles of his particular slavery, he sees that there are others besides him who are enslaved too. He sees that racism is only one dimension of oppression in a whole system of exploitation and racial discrimination, the particular tool of a whole exploitative creed. He sees also that the culture of competition, individualism and elitism that fostered his intellect and gave it a habitation and a name is an accessory to the exploitation of the masses as a whole, and not merely of the blacks. He understands with Gramsci and George Jackson that 'all men are intellectuals'[29] or with Angela Davis that no one is. (If the term means anything it is only as a description of the work one does: the intellect is no more superior to the body than the soul to the intellect.) He realises with Fanon that 'the Negro problem does not resolve into the problem of Negroes living among white men, but rather of Negroes exploited, enslaved, despised by a colonialist, capitalist society that is only accidentally white'.[30] He acknowledges at last that inside every black man there is a working-class man waiting to get out.

In the words of Sartre, 'at a blow the subjective, existential, ethnic notion of blackness[31] passes, as Hegel would say, into the objective, positive, exact notion of the proletariat... "The white symbolises capital as the Negro labour... Beyond the black-skinned men of his race it is the struggle of the world proletariat that he sings" '.[32]

And he sings:

I want to be of your race alone
workers peasants of all lands
...white worker in Detroit black peon in Alabama

uncountable nation in capitalist slavery
destiny ranges us shoulder to shoulder
repudiating the ancient malediction of blood taboos
we roll among the ruins of our solitudes
If the flood is a frontier
we will strip the gully of its endless
covering flow
If the Sierra is a frontier
we will smash the jaws of the volcanoes
upholding the Cordilleras
and the plain will be the parade ground of the dawn
where we regroup our forces sundered
by the deceits of our masters
As the contradictions among the features
creates the harmony of the face
we proclaim the oneness of suffering
and the revolt
of all the peoples on all the face of the earth
    and we mix the mortar of the age of brotherhood
    out of the dust of idols.[33]

Notes

1. Aimé Césaire, in his address to the Congress of Black Writers and Artists, Paris, 1956, reported in 'Princes and Powers', in James Baldwin, *Nobody Knows My Name*, London, 1961.
2. T.S. Eliot, 'The Waste Land', in *Collected Poems*, 1909-1962, London, 1963.
3. The British media uses the 'coloured intellectual', whatever his or her field of work, as white Africa uses the Chief: as a spokesman for his tribe.
4. Jean-Paul Sartre, 'Intellectuals and Revolution: Interview', in *Ramparts*, vol. 9, no. 6, December 1970, pp. 52-5.
5. *Ibid.*
6. Jean-Paul Sartre, *Black Orpheus*, Paris, n.d.
7. Black is here used to symbolise the oppressed, as white the oppressor. Colonial oppression was uniform in its exploitation of the races (black, brown and yellow) making a distinction between them only in the interest of further exploitation — by playing one race against the other and, within each race, one class against the other — generally the Indians against the blacks, the Chinese against the browns, and the coolies against the Indian and Chinese middle class. In time these latter came to occupy, in East Africa and Malaysia for example, a position akin to a

comprador class. Whether it is this historical fact which today makes for their comprador role in British society is not, however, within the scope of this essay. But it is interesting to note how an intermediate colour came to be associated with an intermediate role.
8. Aimé Césaire, *Return to My Native Land*, Harmondsworth, 1969.
9. 'Black: opposite to white.' *Concise Oxford Dictionary*. 'White: morally or spiritually pure or stainless, spotless, innocent. Free from malignity or evil intent, innocent, harmless esp. as opp. to something characterised as *black*.' *Shorter Oxford Dictionary*.
10. Frantz Fanon, *Black Skin: White Masks*, London, 1968.
11. Paul Valéry, quoted in Fanon, *op. cit.*
12. Fanon, *op. cit.*
13. *Ibid.*
14. Thomas Carlyle, 'Occasional Discourse on the Nigger Question', in *Latter Day Pamphlets*, London, n.d.
15. William Blake, from 'Songs of Innocence', in J. Bronowski (ed.), *A Selection of Poems and Letters*, Harmondsworth, 1958.
16. Jean Genet, Introduction to *Soledad Brother: the Prison Letters of George Jackson* by George Jackson, London, 1970.
17. R.D. Laing, *Politics of Experience and Bird of Paradise*, Harmondsworth, 1970.
18. Fanon, *op. cit.*
19. Césaire, *op. cit.*
20. T.S. Eliot, 'Tradition and the Individual Talent' in *The Sacred Wood: Essays on Poetry and Criticism*, London, 1934.
21. *Ibid.*
22. Dunduzu Chisiza, 'The Outlook for Contemporary Africa', in *Journal of Modern African Studies*, Vol. 1, no. 1, March 1963, pp. 25-38.
23. Julius K. Nyerere, *Uhuru na Ujamaa: Freedom and Socialism: a Selection from Writings and Speeches, 1965-67*, Dar-es-Salaam, 1968.
24. *Ibid.*
25. Benjamin Rowland, *Art and Architecture of India: Hindu, Buddhist, Jain*, Harmondsworth, 1970.
26. Paul A. Baran, and Paul M. Sweezy, *Monopoly Capital: an Essay on the American Economic and Social Order*, Harmondsworth, 1968.
27. Frantz Fanon, *The Wretched of the Earth*, London, 1965.
28. He may, of course, become frozen in a narrow cultural nationalism of his own in violent reaction to white culture.
29. Antonio Gramsci, 'The Formation of Intellectuals' in *The Modern Prince and Other Writings*, New York, 1957.
30. Fanon, *Black Skin: White Masks.*
31. 'Negritude' in the original French.
32. Sartre, *Black Orpheus.*
33. Jacques Roumain, quoted in Fanon, *Black Skin: White Masks.*

Part Three

The Black Experience in Britain

Within the space of a few years, from the early 1960s on, the terms of the debate on 'race' in Britain had been set, a common language developed in which that debate was conducted, and its fundamental assumptions established. Blacks were the problem; fewer blacks made for better race relations; immigration control was the answer; social control would follow.

The intellectual backing for these assumptions was provided by the policy-oriented research of the Institute of Race Relations (IRR). The struggle to change the assumptions led to a struggle within the Institute itself and transformed both the Institute and the terms of debate (as told in the pamphlet 'Race and Resistance: the IRR Story', not in this collection).

'Race, Class and the State' emerged from that struggle and the author's own involvement in it — to provide the first coherent class analysis of the black experience in Britain, overturning in the process the old race relations orthodoxies of both right and left. (The occasion was the publication of the White Paper of September 1975 which was to lead to the Race Relations Act of 1976.)

The pieces on the strike of black workers at Grunwick carry further the analysis begun with 'Race, Class and the State', and are an attempt to comprehend and analyse an important development, even as it was unfolding — an analysis which was borne out by subsequent events. And 'From immigration control to "induced repatriation" ', written just before Thatcher's Tory government came into power, warned of the changing nature of racism, from a rationale of exploitation to a rationale for repatriation and predicted the onset of a pass law society (now enshrined in the Nationality Act, 1981).

All these pieces were written in answer to the practical problems facing black struggle. Their theoretical contribution, though influential, was incidental.

Race, Class and the State: The Black Experience in Britain*

For Wesley Dick — poet and prisoner
In some answer to his questions

Within ten years Britain will have solved its 'black problem' — but 'solved' in the sense of having diverted revolutionary aspiration into nationalist achievement, reduced militancy to rhetoric, put protest to profit and, above all, kept a black under-class from bringing to the struggles of the white workers political dimensions peculiar to its own historic battle against capital. All these have been achieved in some considerable measure in the past decade and a half — and the process has already thrown up the class of collaborators so essential to a solution of the next stage of the problem: the political control of a rebellious 'second generation'. And it is to this exercise that the White Paper of 1975 addresses itself.

The political economy of immigration

The laissez-faire era

But to understand the politics of the White Paper, to see what it tells us about state power in one particular aspect — black labour — but an aspect which, like a barium meal, reveals the whole organism of the state and relates black experience to white struggle — one must first reappraise the Immigration Acts. Britain, after the war, like most Western European countries, was faced with a chronic shortage of labour. This shortage was in some measure alleviated by the half a million or so refugees, displaced persons and prisoners of war who were admitted to Britain between 1946 and 1951. But even so, the Ministry of Labour found it necessary to systematise the recruitment of workers from other parts of Europe. Between 1945 and 1957 there was a net immigration of more than 350,000 European nationals into the United Kingdom.[1]

---

* *Race & Class*, vol. XVII, no. 4, Spring 1976.

Unlike most other European countries, however, Britain was in a position to turn to an alternative and comparatively uncompetitive source of labour in its colonies and ex-colonies in Asia and the Caribbean. Colonialism had already underdeveloped these countries and thrown up a reserve army of labour which now waited in readiness to serve the needs of the metropolitan economy. To put it more graphically, colonialism perverts the economy of the colonies to its own ends, drains their wealth into the coffers of the metropolitan country and leaves them at independence with a large labour force and no capital with which to make that labour productive. And it is to these vast and cheap resources of labour that Britain turned in the 1950s.

At first the supply of labour from these countries was governed by the demand for it in the metropolis. Except for a few thousand workers who were recruited directly into London Transport and the British Hotels and Restaurants Association from Barbados (from 1956), no effort was made to relate employment to vacancies. Instead it was left to the free market forces to determine the size of immigration. And this on the whole, as the excellent study by Ceri Peach shows, worked very well.[2] Thus periods of economic expansion led to a rise in immigration, periods of recession to a decline — and this sensitiveness of supply to demand characterised the whole 'stop-go' period of the 1950s.

But if the free market economy decided the numbers of immigrants, economic growth and the colonial legacy determined the nature of the work they were put to. It was inevitable that in a period of full employment the indigenous worker would move upwards into better paid jobs, skilled apprenticeships, training programmes, etc., leaving the dirty, hard, low-paid work to immigrant labour. Although, that is, the shortage of labour was general, the more dynamic and attractive sectors of industry were able to draw the best qualified labour from both the non-growth industries as well as the immigrant labour force. The non-growth sector (including the public services), on the other hand, had only new entrants to the labour market to turn to. (In practice, though, prejudice decreed that qualified immigrants were more available to the latter than to the former.) Thus the jobs which 'coloured immigrants' found themselves in were the largely unskilled and low status ones for which white labour was unavailable or which white workers were unwilling to fill — in the textile and clothing industries, engineering and foundry works, transport and communication, or as waiters, porters, kitchen hands.

And since the opportunities for such work obtained chiefly in the already overcrowded conurbations, immigrants came to occupy some of the worst housing in the country. The situation was further exacerbated by the exhorbitant rents charged by slum landlords. Attempts on the part of the newcomers to break the landlords' hold by buying their own homes were often frustrated either by the difficulties of obtaining loans from regular sources or by the prohibitive rates of interest charged by the irregular ones — or even by the refusal of owners to sell to 'wogs' and 'nig-nogs'. When immigrants eventually managed to buy their own property and were able to house their fellows, they were accused of overcrowding — sometimes sleeping five and ten to a room. (That there was excellent precedent for this in the dormitories of Eton and Harrow went unnoticed and unremarked.) In the course of time the 'immigrants' became ghetto-ised and locked into the decaying areas of the inner city. And a ghetto, in the words of Ceri Peach, 'is the geographical expression of complete social rejection'.[3]

Everyone made money on the immigrant worker — from the big-time capitalist to the slum landlord — from exploiting his labour, his colour, his customs, his culture. He himself had cost the country nothing. He had been paid for by the country of his origin — reared and raised, as capitalist under-development had willed it, for the labour markets of Europe. If anything, he represented a saving for Britain of all the expense involved in feeding and clothing and housing him till he had come of working age. For, as André Gorz has pointed out, 'the import of "ready-made" workers amounts to a saving, for the country of immigration, of between £8000 and £16000 per migrant worker, if the social cost of a man is estimated for Western European countries as between five and ten years of work'.[4] And the fact that in the early years of migration, the 'coloured' worker came to Britain as a single man — as a unit of labour — unaccompanied by his family meant an additional saving to the country in terms of social capital: schools, housing, hospitals, transport and other infra-structural facilities. A fraction of the saving made from the import of these ready-made workers — let alone their active contribution in labour and taxes — could have served to increase social stock and improve social conditions if the government had so willed. But capital and the state were concerned with the maximisation of profit, not with the alleviation of social need.

By the late 1950s, however, the contradiction between the social and economic needs of Britain, *thrown up* — not caused — by immigration, became more defined. The shortage of workers, as Ceri Peach shows, made immigrants economically acceptable; the shortage of housing made them socially undesirable. 'The colour prejudice of landlords and landladies coupled with the shortage of houses made the crowding, and in some cases the overcrowding, of much of the accommodation available to the migrants inevitable and this, in turn increased their image of undesirability.' From being refused accommodation on the grounds that they were coloured, they were now refused houses on the grounds that they would overcrowd. 'It is surely an ideal system,' concludes Peach, 'in which prediction produces its own justification.'[5]

Ideal, that is, for capital — for it gets labour without the overheads (so to speak), profit without pain, gain without cost. Having already deprived one section of the working class (the indigenous) of its basic needs, it now deprives it further in order to exploit another section (the blacks) even more — but, at the same time, prevents them both from coming to a common consciousness of class by intruding that other consciousness of race. It prevents, in other words, the horizontal conflict of classes through the vertical integration of race — and, in the process, exploits both race and class at once.

To put it differently, the profit from immigrant labour had not benefited the whole of society but only certain sections of it (including some sections of the white working class) whereas the infrastructural 'cost' of immigrant labour had been borne by those in greatest need. That is not to say that immigrants (qua immigrants) had caused social problems — Britain, after all, was a country of net emigration — but that the *forced* concentration of immigrants in the deprived and decaying areas of the big cities high-lighted (and reinforced) existing social deprivation; racism defined them as its cause. To put it crudely, the economic profit from immigration had gone to capital, the social cost had gone to labour, but the resulting conflict between the two had been mediated by a common 'ideology' of racism.

Prelude to control

That same 'ideology' detonated the race riots of 1958 — and revealed to the state that considerations of social need had now to be weighed against considerations of economic gain. Racism, though economically useful, was becoming socially counter-productive. And the state, which had hitherto acted in the economic interests of the ruling class, was now compelled to modify that role and assume its other function of appearing to act in the interests of society as a whole — in the 'national interest'. The first step was to slow down immigration, thin out the black presence, the second to manage racism, keep it within profitable proportions — relief for the depressed areas, urban aid, would follow. The economy in any case had, for the time being, absorbed all the unskilled labour it could (though it still required skilled and professional workers). Additional units of labour applied to existing (outworn, outmoded) plant would not yield the returns that would make such addition justifiable. On the other hand, automation and new technology — capital intensive production — would help Britain to compete with the rest of Europe in markets made more competitive by the loss of its colonies. That same 'loss', however, would make it possible for Britain to renege on its Commonwealth ties and look to the Common Market for the labour it required — when the time was ripe. The stage was set for immigration control.

To end immigration altogether would have been one answer. But given the periodic labour shortages characteristic of the capitalist countries of Western Europe, given the structural needs of late capitalism for the import of foreign workers, it was no answer at all. Migrant labour, precisely because it was migrant — seasonal and contractual, filling in the labour gaps in times of expansion and being fired in times of recession — served to absorb the shocks of alternating booms and depressions. And by virtue of the fact that it was foreign, migrant labour yielded extra-profit to the employer.[6] Most of Western Europe had worked out a migratory mechanism combining both these functions. Labour, on short-term permits, on contract, ensured the buffer function; and the fact that it was foreign, recruited from the underdeveloped southern extremities of Europe, ensured that it would not — by virtue of nationality laws freely agreed to — have the same rights as the indigenous worker and could therefore be discriminated against. And to discriminate is to exploit, to derive a surplus value larger than that afforded by the exploitation of the native worker.[7] Together they, contract labour and nationality laws, fulfilled a third function — a political one: they prevented the integration of migrant labour into the indigenous proletariat and thereby mediated class conflict.

Britain, still outside the European community but periodically knocking at its door and gifted with a vast reserve of labour in the colonies and Commonwealth, was loth to let go of either and tried to hang on to both. Initially it recruited migrant workers from Europe on a permit basis. Between 1946 and 1951, 100,000 European workers had entered Britain. But the availability of labour in the colonies and ex-colonies and its sensitivity to demand made labour on contract unnecessary.[8] And as for a discriminatory mechanism, in place of nationality laws there was the fact of race. Black labour was inherently 'discriminatable'. It was alien per se — and automatically excluded from integration into a racist white working class.

It had suited Britain, therefore, to import the workers it needed from its colonies and ex-colonies: it was the quickest way of getting the cheapest labour at minimum (infrastructural) cost — and without the fuss and bother of barriers. It worked, in effect, like any internal migratory movement: a movement of population from the periphery to the centre as and when the need arose. And in that sense it was unrestrained, laissez-faire. But to characterise the laissez-faire period of immigration as an essay in British absent-mindedness — the sort of aristocratic whimsy that gathers and loses empires on the spin of a wheel — or as a conscious 'open-door' policy designed to benefit the poor orphaned children of empire as befitted a once and only mother country — an aspect of British high-mindedness — is a load of bull-shit.[9] So ingrained were these views among radical analysts that when, over the 'Kenyan Asian' affair (in 1968), Labour went even more Tory than Tory, the 'experts' instead of abandoning their analysis, mourned instead the death of Labour idealism or, more concretely, the passing of 'the liberal hour' — and of Roy Jenkins, its finest flower.

The fact of the matter was that laissez-faire immigration and laissez-faire discrimination had thrown up social problems which, after the riots of 1958 and the growing militancy of a black under-class were taking on political proportions that the government — irrespective of party — could not ignore. It had to put an end to 'coloured immigration' and yet have recourse to a reserve pool of labour when required. The crux of the problem, therefore, was not migration, but settlement — and not discrimination but *racial* discrimination. For the purposes of exploitation, it was labour and not colour that had to be discriminated against — and that could be done on the basis of citizenship, of nationality, rather than of race. And since nationality laws by definition distinguished between citizen and alien, foreign or migrant labour would be automatically subject to discrimination. To change British nationality laws so as to put Commonwealth citizens on a par with aliens was the most obvious solution — and it had the added advantage of debarring settlement as a matter of right. But, on the other had, it would spell the end of a historical relationship which ensured the continuing dependency of the colonial periphery on the centre. (No one, bar the tear-stained liberals, believed the sentimental bull about mother-country obligations.) The aim, therefore, was to move gradually towards the European model of contract labour (and a European configuration with the poor south as its periphery) without foregoing the 'Commonwealth' relationship. Eventually the Commonwealth relationship would have to be subordinated to the European relationship — and then the nationality laws would need to be tidied up — but for the time being a solution had to be found that did not require such a change.

This meant, in concrete terms, that immigrants from the Commonwealth countries, though remaining British subjects under British nationality law, would be debarred from entering (and settling in) Britain except as and when required by the British economy. Thus the formal links with the Commonwealth would be maintained but the right of individual citizens to automatic entry would be denied. In terms of British nationality law, this would mean that a British citizen was not completely a British citizen when he was a black British citizen — somewhat on the lines of the American constitution which once decreed that a 'Negro' was three-fifths of a person. Nevertheless it would be a solution to black settler immigration: if it did not end settlement altogether it would at least reduce the numbers.

From status to contract

Accordingly the Commonwealth Immigrants Act of 1962 restricted the admission of Commonwealth immigrants for settlement to those who had been issued with employment vouchers. The vouchers themselves were chiefly available to those who had jobs to come to (A vouchers) and to those with skills and qualifications 'likely to be useful in this country' (B vouchers). A third category, C vouchers, for unskilled workers gradually disappeared and became a dead letter by September 1964.[10] And, as though to compensate for the discrimination now institutionalised in the Immigration Act, a Commonwealth Immigrants Advisory Council (CIAC) was set up to advise the Home Secretary on immigrant welfare and integration.

When the Labour government came to renew the Commonwealth Immigrants Act in the White Paper of August 1965, it made further restrictions on 'coloured' immigration — reducing the number of vouchers in the A and B categories to a ceiling of 8,500 per year and doing away with the unskilled category C altogether. It also reduced the categories of skill and qualifications required of B voucher applicants to doctors, dentists, nurses, teachers and graduates in science and technology. The policy was now firmly established that immigration from the black Commonwealth should be geared to the requirements of the British economy.[11] And since the manpower needs of this period were infrastructural — the schools (including medical schools), hospitals, houses, etc., that the state had decided not to invest in during an earlier period — it was to the skilled and the professional that employment vouchers were increasingly issued. Over 75 per cent of the vouchers issued in the first half of 1966 alone were to such personnel, whereas for the whole of 1965 the figure was 55 per cent. Or take another statistic: of the 3,976 B vouchers[12] (A vouchers accounted for 306) issued to India in 1966, 1,511 went to doctors, 922 to technology graduates, 667 to teachers and 469 to science graduates (and 407 to others).[13] Any lingering pretence that the employment of Commonwealth immigrants aided the Commonwealth was dispelled by a system which creamed off the most skilled and professional personnel from these countries while keeping out their unskilled.

It was also a system which took discrimination out of the market place and gave it the sanction of the state. It made racism respectable and clinical by institutionalising it. But in so doing it also increased the social and political consequences of racism. And to counter these the state set out to develop a more coherent policy of integration. Thus, the White Paper replaced the CIAC with the National Committee for Commonwealth Immigrants (NCCI). In fact, the announcement of other legislation to deal with 'racial discrimination in public places and with the evil of incitement to racial hatred' preceded the White Paper. But an examination of the politics of integration (as opposed to the sociology of integration) belongs to the second half of this paper. Here it is intended to pursue the investigation into immigration policies to see how they effected the transition of Commonwealth (and therefore British) citizens from the status of citizens to labourers on contract.

The Commonwealth Immigrants Act of 1968 is not essential to that investigation — except in that the circumstances which necessitated its enactment highlighted yet more the contradiction between British nationality laws and the Immigrants Acts and once again pointed to the passage of the Commonwealth citizen from status to contract.

In 1967, following on the Africanisation policies of the Kenyatta government, British Asians in Kenya, who had not opted for Kenyan citizenship at independence (1963) and had stayed loyal to the 'mother country', were granted only temporary residence. They were in effect asked to go home to Britain. Already in 1965 and 1966 six thousand Asians, possessing

British citizenship, who were not subject to immigration control, had entered the UK. But after the Kenyan legislation of 1967, the numbers increased and the British (Labour) government, with an eye to all those other British Asians and British Chinese whom Britain had used and abandoned on the darker shores of the once empire, decided that they were not as British as their passports warranted. They were only as British as Commonwealth citizens. And since they were liable to the voucher system, the British Asians in Kenya would also be liable to the same procedure for admission — but would be allocated special vouchers as distinct from work vouchers.

Given the devaluation of British citizenship in 1962, the distinction between Commonwealth citizens and Kenyan British Asians was only a legal nuance — except that, unlike the former, the Kenyan Asians had nowhere but Britain to go to: they were potentially stateless. And this aspect plus the fact that they were more middle-class and British than the normal run of immigrants particularly outraged British liberal opinion. But the blacks, post-1962, had seen the Act merely as the correction of an anomaly in the policy of reducing all black British citizens to the lowest common denominator of contract labour.

As usual, new anti-discriminatory legislation and integrationist policies went hand in hand with the new Immigrants Act — but these again will be dealt with in the next section.

The 1962 and 1965 Immigrants Acts had ensured the supply of skilled and professional workers from the black Commonwealth; for the seasonal unskilled jobs Britain turned to 'foreign workers'. The Kenyan Asian episode had temporarily swelled the number of settlers beyond immediate employment needs (the voucher system, in this case, was a device to phase-in the Kenyan 'exodus'). And the 1968 Act had in effect brought 'coloured' UK passport-holders within the provisions of the Immigration Acts. Black settler migration was firmly under control, but it was still settler and not migrant. Once in, the black 'immigrant' could remain in the UK indefinitely — and after five years he had the right to British citizenship. He was still not a fully-fledged '*Gastarbeiter*'.

That situation was remedied by the Immigration Act of 1971 which put him, finally, on the same footing as the foreign worker:

he could only come in on a permit to do a specific job in a specific place for an initial period of not longer than twelve months. He could not change his job without the permission of the government — which meant that he was dependent on his employer for recommendation: he had to be a good little wage-slave. He may, like any other alien, apply for UK citizenship at the end of four years, provided that he has been 'of good behaviour'. On the other hand, he could, if the Home Secretary so wished, be deported on the ground that it was 'conducive to the public good as being in the interest of national security or of the relations between the UK and any other country or for other reasons of a political nature'.

The immigrant was finally a migrant, the citizen an alien. There is no such thing as a 'Commonwealth immigrant' anymore. There are those who came from the Commonwealth before the 1971 Act came into force (January 1973) but these are not immigrants; they are settlers, black settlers. There are others who have come after the Act; they are neither settlers nor immigrants, they are simply migrant workers, black migrant workers. And the migratory mechanism — the combination of contract labour and discriminatory nationality laws — which ensures that the *Gastarbeiters* of Europe are no more than second-class production factors yielding surplus surplus value as well as acting as a buffer, a shock absorber, between boom and depression now applied to migrant workers from the 'Commonwealth' except that time and distance and fares and race made them less accessible to the British labour market than their European counterpart. Then there are the workers, since Britain's entry into Europe in 1975, from the European *community* with free access to work in Britain. And there are aliens and colonials and patrials and non-patrials and white Commonwealth... All of which makes a mess of nationality laws and discrimination less tidy — and for those reasons must claim the government's attention in the near future. But all of which also leaves the divisions and sub-divisions within the non-indigenous sector of the working-class — apart from the divisions between them, a sub-proletariat, and the native workers — looking something like this:

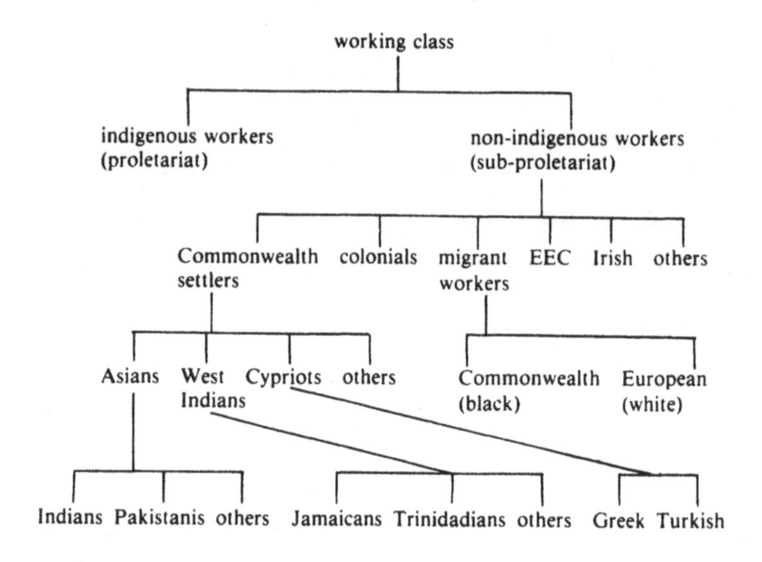

Britain now had two main reserve pools of labour: in the under-developed south of Europe and in the under-developed Third World — one for unskilled and/or seasonal labour, the other for skilled and professional — one, to put it crudely, to do the dirty work, the other to provide the infrastructural facilities (transport, hospitals, welfare) needed to keep the workers working — but neither exclusively so.[14] In a sense, Britain was now a neo-colonial power with two peripheries. And if migrant labour helped to perpetuate both these dependencies, the older was also anchored in that other history of colonialism.

The Politics of Integration

From institutional racism...

Thus the state had achieved for capital the best combination of factors for the exploitation of labour while appearing, at the same time, to have barricaded the nation against the intrusion of an 'alien wedge'. It had atomised the working class and created hierarchies within it based on race and nationality to make conflicting sectional interests assume greater significance than the interest of the class as a whole. It had combined with the trade union aristocracy to reduce the political struggle of the labour movement to its bare economic essentials — degraded the struggle to overthrow the system to the struggle to be well off within it — and in the process had weaned the trade unions from the concerns of the labour movement to the concerns of government. And when the black proletariat threatened to bring a political dimension, from out of their own historic struggle against capital, to the struggles of the working class, state policy had helped trade unions to institutionalise divisive racist practices within the labour movement itself.

But racism is not its own justification.[15] It is necessary only for the purpose of exploitation: you discriminate in order to exploit or, which is the same thing, you exploit by discriminating. So that any other system of discrimination, say on the basis of nationality, would — if available — do equally well. During the laissez-faire period of immigration, racism helped capital to make extra profit off black workers (extra in comparison to indigenous workers) — and the state, in the immediate economic interests of the ruling class, was content to leave well enough alone. But in the 1960s the state, in the long term and overall interests of capital (as against its temporary and/or sectional interests), entered into the task of converting immigrant settler labour to migrant contract labour. One of the benefits of such labour, as has been shown, is that it is automatically subject to discrimination on the basis of nationality laws and inter-state agreements. The British government, however, had — for reasons outlined earlier — no wish to change the nationality laws in order to stop 'coloured immigration' — some of the Caribbean countries were still colonies anyway.[16] Hence it resorted to a system of control which, in being specifically (though not overtly) directed against the 'coloured' Commonwealth, was essentially racist.

The basic intention of the government, one might say, was to anchor in legislation an institutionalised system of discrimination against foreign labour, but because the labour happened to be black, it ended up by institutionalising racism instead. Instead of institutionalising discrimination against labour it institutionalised discrimination against a whole people, irrespective of class. In trying to banish racism to the gates, it had confirmed it within the city walls.

The whole thing was particularly 'untoward' because once immigration control had helped to minimise the number of blacks settling in Britain, the 'black problem' itself would have become more manageable. And the lessons of America had not been lost on Britain. Hence in order to counter-act the consequences of the Immigration Acts and to stop black militancy from infecting the body politic, the government embarked on a programme of 'integration'.

The Commonwealth Immigrants Advisory Council of 1962, however, was no more than a gesture towards integration: its function was to advise the Home Secretary on matters of immigrant welfare. But with the White Paper of 1965 integration began to assume the proportions of a philosophy. In fact the government had, in introducing further controls on immigration, pointed out that the purpose of reducing the numbers coming in was to improve matters for those already within — to improve race relations. 'Without integration,' opined a future minister, 'limitation is inexcusable; without limitation, integration is impossible.'[17] Accordingly the government replaced the CIAC with the National Committee for Commonwealth Immigrants (NCCI), with lots of money and staff and local liaison committees — and, to vest the effort with sanctity, set the Archbishop of Canterbury at its head. It was an independent body, however, free of government control, but linked to it through a minister in the Home Office with 'special responsibility for immigrants'. The Committee's brief was 'to provide and coordinate on a national basis efforts directed towards the integration of Commonwealth immigrants into the community'.

The government also introduced the first piece of antidiscriminatory legislation in the form of the Race Relations Act of 1965, but this was a half-hearted affair which merely forbade discrimination in 'places of public resort' and, by default, encouraged discrimination in everything else: housing, employment, etc. The incorporation, in the Act, of a clause to 'penalise incitement to racial hatred' turned out to be more useful in imprisoning blacks (and right-wing extremists) than in arresting the exalted nativism of the Rt. Hon. Enoch Powell, Ronald Bell Q.C. and others of their ilk and silk. The discrimination provisions of the Race Relations Act were to be implemented by the Race Relations Board and its local conciliation committees.

But the concern of integration during this period related more to the Asians than to the West Indians. The latter, it was felt, had 'largely been brought up to regard themselves as British', whereas 'Pakistanis and Indians... showed almost no interest in being integrated'.[18] The Asians, with their different cultures and customs and language and dress, their extended families and sense of community, and their peculiar preference to stay with their own kind, were a society apart. But they were also a people who were industrious and responsible, anxious to educate themselves, prepared to work hard and move up the social and economic ladder, honest, diligent, 'politic, cautious and meticulous' — all virtues which shored up bourgeois society. Besides they too had their creeds and their castes and their classes. They may not be assimilable, but they were certainly made for integration — a parallel society to be accommodated in a pluralist set-up. All that was required was an acceptance of the principles of cultural pluralism on the part of the 'host' population. And it was not as though Britain had not had a tradition of accommodating other cultures and other peoples — only, this time, they happened to be a little more different and a lot more visible. Hence the precision with which the Home Secretary, Roy Jenkins, defined integration in May 1966: 'not as a flattening process of assimilation but as equal opportunity accompanied by cultural diversity, in an atmosphere of mutual tolerance'.[19]

The West Indians, on the other hand — it had been assumed — were a part of British culture, an aspect of it, a sub-culture. They spoke the same language, wore the same clothes, followed the same religions. They were not a society apart — only their colour was different. They could be assimilated into the mainstream of British society.[20] All that was necessary to make them acceptable to the 'host' society was to banish colour prejudice, outlaw racial discrimination.

The NCCI and Race Relations Board, however, did not succeed in even getting that programme off the ground. The Board was virtually a non-starter, so feeble and narrow were the provisions of the 1965 Act. The National Committee discovered discrimination everywhere it went but was frustrated into educating people out of their attitudes. 'Education in school and out of school, education of adults as well as children, education of newcomers as well as the indigenous population, education through conferences, through committee work, through social activities, through the Press...' dragged on its first annual report in the tones of a forlorn manifesto. Hence in 1966 both bodies jointly commissioned the PEP (Political and Economic Planning) to investigate the extent of racial discrimination. Its report, published a year later, produced evidence to show what everybody knew: that racial discrimination varied in extent from 'the massive to the substantial.'

The profound effects of racism were already showing in the growing militancy of the West Indian community. At first it was only civil rights and Martin Luther King that had claimed their attention. The Campaign Against Racial Discrimination (CARD), set up in December 1964 after Martin Luther King's visit to London in transit to Sweden to receive the Nobel prize for peace, was composed of West Indians (militant and 'normal'), Asians (mostly 'normal') and whites (liberals and radicals). Its task was to fight discrimination by lobbying Whitehall, by asking the government — but in tones so strident as to pass for passion — to be nice to the blacks. But increasing police harrassment, particularly of West Indians, mounting discrimination in employment and housing and the relegation of West Indian children to ESN (Educationally Sub-Normal) schools sparked off militant struggles in the Caribbean community. The black rebellion in America gave fillip to black nationalism. And Stokely Carmichael's visit in 1967 signalled the high water mark of revolutionary black politics. CARD, like the proverbial house of, folded under the impact, leaving it to *The Times* news team to conclude that 'the ominous lesson of CARD... is that the mixture of pro-Chinese communism and American-style Black Power on the immigrant scene can be devastating'.[21]

The state was faced, against all its convictions, with an unassimilable black community. The West Indians were not a part of British society after all. They even proclaimed that they had a culture and a tradition and a history of their own. They rejected British values and British culture. And worse, especially for the educationalists who had suddenly come upon the discovery that the West Indian child could not/would not speak English English: they rejected the English language itself. Once, as slaves, when they had been forced to accept the white man's language, they had corrupted it so skilfully as to make it unintelligible to the slave-master. Now they sought to 'blacken the language, suffuse it with their own darkness and liberate it from the presence of the oppressor'.[22] And out of that assertion of themselves was springing an anti-capitalist ideology and a politics of revolution. They posed a problem from within British society — they posed the problems of it. They could not be assimilated and they could not, like the Asians, be integrated. They were a canker in the body politic. The body politic itself was threatened. The need for integration and for anti-discriminatory legislation had assumed a new urgency.

But as the government was contemplating fresh legislation, the 'Kenyan Asian' storm broke — and led to the Commonwealth Immigrants Act of March 1968. In April Enoch Powell warned his people that 'their wives [were] unable to obtain hospital beds on childbirth, their children unable to obtain school-places, their homes and neighbourhoods changed beyond recognition'. In May the Prime Minister, Harold Wilson, responded by promising an urban programme which would give substantial aid to local authorities 'in special need'. In October the Race Relations Act became law.

The 1968 Act extended the scope of the 1965 Act to include discrimination in employment (with some exceptions), housing (with some exceptions), credit and insurance facilities and places of public resort. But the breadth of its concerns was belied by the unenforceability of its provisions. The Board would have to rely almost entirely on conciliation to obtain redress. It had no powers of enforcement but could resort to the courts, in extreme cases, to obtain an injunction restraining the defendant from further discriminatory practices. It could order the payment of special damages and damages for the loss of opportunity.[23] All penny-pinching stuff. Basically the Act was not an act but an attitude.

But then it was never meant to be anything else. Anti-discriminatory legislation was not meant to chastise the wicked or to effect justice for the blacks. If it was, the government would have had no difficulty in making its intention felt in the administration of the law. Its sole purpose, however, was education — the education of the lesser capitalists in the ways of enlightened capital. Racial discrimination was a short-term expedient to exploit a section of the working class, and now that immigration laws were turning immigrants into migrants — and migrants from Europe would soon become available — it was necessary to count the social and political cost of racial friction.

...to domestic neocolonialism

The purpose of the Board as far as the state was concerned was to carry that lesson to employers and local officials. And it was a lesson to be taught not in anger or in punishment but in sorrow and conciliation. The very structure and personnel of the Board and its conciliation committees, marked by the presence of local firms and interests (and token blacks) and the absence of black workers from the factory-floor, bear witness to the point and purpose of the Act.

And yet there have been protestations that the Board has failed. Failed for the masses of the blacks, yes. But it succeeded in what the state meant it to do: to justify the ways of the state to local and sectional interests — and to create, in the process, a class of coloured collaborators who would in time justify the ways of the state to the blacks. One has only to look at the successful cases handled by the Board to see how much of it relates to the alleviation of marginal and often middle-class discrimination. In the year April 1969 to March 1970, for instance, the highest percentage of success recorded related to discrimination in clubs: 50 per cent, the lowest to dismissals in employment: 4 per cent. Or take a look at the Boards's journal, *Equals*, not just for the 'black' columnists who rage on the page they are paid for, but to see how the blacks are making it in the system. Sewa Singh Sodi can now play darts for his local against a club that once operated a colour bar. Mr Trevor MacDonald is 'the first black staff reporter to present news on British national TV', Mr Yunus Chowdry is the first black man to reach the National

Executive of his union, the National Union of Dyers, Bleachers and Textile Workers, a veritable trade union aristocrat with his own little fiefdom. There is also a first black mayor somewhere in Wales, a first black woman deputy mayor in Camden, and of course there is the black black Lord, Lord Pitt — not the first, but close. And all within the last couple of years.

The 1968 Act also re-formed the existing organisation, the NCCI, to create the Community Relations Commission — in order to complement the work of the Race Relations Board. The Commission's task as defined by the Act was to 'promote harmonious community relations', to co-ordinate national action to this end through its local community relations councils, to disseminate information about matters affecting minority groups and to advise the Home Secretary.

In theory, the Commission attempted to combat racial discrimination, the Board to penalise it. In practice, they were both educational and advisory and tended to overlap each other. In effect, they were to one degree or another both instruments of mediation — between sections of the ruling class, between the sectional interests and the blacks and, on the national level, between whites and blacks.

In its seven years of existence, the Commission has succeeded in saturating the key areas of society with information, advice and literature explaining West Indian and Asian peoples to white groups and individuals in positions of influence and power — employers, police, political parties, churches, local authorities, voluntary groups, educationalists, trade unions, the media. It has held conferences and seminars, often jointly (but in an elder capacity by virtue of its specialised and/or statutory position) with bodies such as the National Union of Journalists, Inner London Education Authority, Library Association and Department of Education and Science — to inform the future holders of power: trainees in youth work, community work, teaching, policing, etc. Only in matters of employment and labour relations and the sophistry of statistics has it seemed to rely on the efforts of that 'independent' body, the Runnymede Trust.

In the structure of its local community relations councils, the Commission revealed the success of its local, grassroots effort. In the main appointed by the statutory body, the Commission, and paid by the local authority, but always governed by Councils which are an exact replica of the local power structure (businessmen, police, political parties, trade unions, headmasters, clergy), the office of the Community Relations Officer defines, exactly, 'integration' as the absorption and negation of black discontent: the accommodation within the local status quo of factors that threaten the status quo, the expansion of the status quo itself to accommodate such factors.

But most important of all, the Commission took up the black cause and killed it. With the help of its 'black' staff and its 'black' experts, with the help of an old colonial elite and through the creation of a new one, it financed, assisted and helped to set up black self-help groups, youth clubs, supplementary schools, cultural centres, homes and hostels. It defined and ordained black studies; it investigated black curricula; it gave a name and a habitation to black rhetoric. And finally, almost in a last blaze of glory, the Commission, funded for this purpose by the Gulbenkian Foundation, brought together at a residential conference in an opulent hotel in January 1975 a cross-section of black activists, gave up the platform to the most militant blacks and itself sat in the aisles, servicing the black people. Aptly, in view of the new dependent relationship that the black community was entering into, the Conference was named 'Black People: The Way Forward'. And out of that conference has emerged a new black committee one of whose functions will be to advise the Gulbenkian Foundation (shades of Ford) where to put its money — in itself an indication that the black programmes can now be safely left in private enterprise.

The Commission's task is over. The Race Relations Bill (February 1976) sees that its work is good and that its work is done. It has taught the white power structure to accept the blacks and it has taught the blacks to accept the white power structure. It has successfully taken politics out of the black struggle and returned it to rhetoric and nationalism on the one hand and to the state on the other. It has, together with the Board, created a black bourgeoisie, especially West Indians (the Asian bourgeoisie was already in the wings), to which the state can now hand over control of black dissidents in general and black youth in particular. Britain has moved from institutional racism to domestic neo-colonialism.

In terms of the larger picture, what has been achieved in half a decade is the incorporation of a whole generation of West Indian militants. The Asians had already settled into the cultural pluralist set-up ordained for them by the state as far back as a decade ago. They had their own TV and radio programmes, their mosques and their temples, their shops and cinemas and social centres. More importantly, they had thrown up leaders and spokesmen who spoke to and worked with the state. They had remained parallel in terms of culture, they had merged in terms of class. Only in regard to the Asian working class was there any trouble. Their strikes at Courtauld's and Woolf's and Mansfield Hosiery and Imperial Typewriters had threatened the system as few strikes did — for they were subsidised and supported by the community, united across divisions of labour and possessed of a genius for organisation and obstinacy against all sorts of odds (including trade union ones).

The strategy of the state in relation to the Asians had been to turn cultural antagonism into cultural pluralism — in relation to the West Indians, to turn political antagonism into political pluralism. The process, in the case of the Asians, was first to free them from cultural oppression so as to help them 'modernise' their own class hierarchies and social structure — and then slot them into mainstream society. The West Indians, however, had to undergo a different process — for they were an aspect of British culture and society and yet outside it, even antithetical to it. Their similarities might have arisen from a master-slave history into which they had been locked in deathly embrace, but that same history had produced a culture and a politics that were mortally anti-white and anti-capital. The task, therefore, was to separate their antagonisms: to leave them anti-white but make them pro-capital. The task was to free them from the dishonour of racism so that they could honour the blandishments of capital. They had to be allowed to move upwards within the existing system so that they would not threaten to transform it into a different system.

But there was still 'the second generation'. All the other blacks had been found a place within the system, but the young blacks stood outside it. As though to confirm the dialectics of history they, the British born, carry the politics of their slave ancestry. And so it is to them that the state now turns its attention in the Race Relations Bill of February 1976 and the White Paper that heralded it.

The Politics of the White Paper

Listen to the voice, the anxieties of the state:

> the character of the coloured population resident in this country has changed dramatically over the decade. Ten years ago, less than a quarter of the coloured population had been born here: more than three out of every four coloured persons then were immigrants to this country... About two out of every five of the coloured people in this country now were born here and the time is not far off when the majority of the coloured population will be British born.

Some of these the state has already mobilised by affording them places in universities and colleges of higher education, others it has taken care of — in borstals, mental homes and prisons. But some, in a completely unprecedented new phenomenon, have picked up the gun — not of course in the organised manner of a revolutionary political party or even as a movement (for as one small fragment of a very small minority, black youth qua black youth cannot have a mass base) but as self-ordained soldiers of the people.[24] That is not to romanticise their futile ambition to lay siege to the state but to acknowledge, even while acknowleging the romanticism of the act, the deep dark concern out of which their commitment springs. It is to acknowledge their gesture as a new language of resistance — and to refute the definition which the state through years of indoctrination has persuaded the black under-class to accept as the language of gangsterism. It is to refute, in the particular, that other romanticism of anti-organisation blacks which holds that unemployed black youth or, rather, anti-employment black youth are 'gunning for a wage'.[25] It is, in other words, to refute the 'ideology' of these political romanticists that if every dissident section of society did its own thing, capitalism would lie down and die — it is to refute the politics of spontaneism which Gramsci equates with opportunism.

And it is to assert that 'the union of spontaneity and conscious leadership, or discipline *is* the real political action of subaltern classes, in so far as this is mass politics and not merely an adventure by groups claiming to represent the masses'.[26]

For, the anxiety of the state about rebellious black youth stems not from the rhetoric of professional black militants (whose dissidence it can accommodate and legitimise within the system) but from the fear of the mass politics that it may generate in the black under-class and in that other discriminated minority the migrant workers and perhaps in the working class as a whole — particularly in a time of massive unemployment and urban decay.

Almost a decade earlier the Home Secretary had warned the country against the future depredations of the second generation and argued for timely attempts at 'civilised living and social cohesion'. Now in the White Paper the same Home Secretary pointed out the 'politically grave consequences' of continued racism.

> If... job opportunities, educational facilities, housing and environmental conditions are all poor, the next generation will grow up less well-equipped to deal with the difficulties facing them. The wheel then comes full circle, as the second generation find themselves trapped in poor jobs and poor housing. If, at each stage of this process an element of racial discrimination enters in, then an entire group of people are launched on a vicious downward spiral of deprivation.

Thus the Bill's intention is not just to outlaw discrimination but to carry the fight against discrimination into every area of society — housing, education, employment, trade unions, local government, vocational training bodies, etc. And more significantly, it means to *enforce* the law. The law is no longer an instrument of education; it is an instrument of compulsion. More, it will redress the balance of discrimination in some areas by discriminating in favour of the disadvantaged blacks; for it acknowledges at last that although 'they may share each of the disadvantages with some other deprived group in society... few other groups in society display all their accumulated disadvantages'. 'It is no longer necessary to recite the immense danger, material as well as moral, which ensues when a minority loses faith in the capacity of social institutions to be impartial and fair.' And that is why the government believes that 'it is vital to our well-being as a society to tap these reservoirs of resilience, initiative and vigour in the racial minority groups and not allow them to lie unused or to be deflected into negative protest on account of arbitrary and unfair discriminatory practices'.

Hence the new Race Relations Commission which will replace both the Board and the Commission 'will have a major *strategic* role in *enforcing* the law in the *public interest*' (emphasis added).

However that interest is defined — as 'the public interest' or the national interest or, unashamedly, the ruling class interest — it is certainly the interest of capital. For capital requires racism not for racism's sake but for the sake of capital. Hence at a certain level of economic activity (witness the colonies) it finds it more profitable to abandon the idea of superiority of race in order to promote the idea of the superiority of capital. Racism dies in order that capital might survive.

Notes

1. See Stephen Castles and Godula Kosack, *Immigrant Workers and the Class Structure in Western Europe*, London: OUP for IRR, 1973, and E.J.B. Rose and others, *Colour and Citizenship: a Report on British Race Relations*, London: OUP for IRR, 1969.
2. Ceri Peach, *West Indian Migration to Britain: a Social Geography*, London: OUP for IRR, 1969.
3. *Ibid.*
4. André Gorz, 'The Role of Immigrant Labour', *New Left Review*, No. 61, May-June 1970.
5. Peach, *op. cit.*
6. See M. Nikolinakos, 'Germany: the Economics of Discrimination', *Race Today*, Vol. 3, no. 11, November 1971, pp. 372-74
7. See A. Sivanandan, 'Race, Class and Power: an Outline for Study', *Race*, Vol. 14, no. 4, April 1973; pp. 383-91.
8. '...The rate of immigrants into this country,' said Gaitskell to the Commons in the debate on the 1962 Act, 'is closely related and, in my view at any rate, will always be closely related to the rate of economic absorption... There has been over the years... an almost precise correlation between the movement in the number of unfilled vacancies, that is

to say employers wanting labour, and immigration figures' — House of Commons Official Report, Vol 649, Col. 793, 16 November 1961.

9. 'Through the 1950s Britain acquired a coloured population in, so to speak, a fit of absence of mind.' Dipak Nandy in the Foreword to *The Multi-Racial School*, by Julia McNeal and Margaret Rogers, Harmondsworth, Penguin, 1971. For the high-minded school, see, for instance, the writings of P. Mason *et al.*

10. It is significant that at the time the Immigration Bill was being debated Britain was negotiating for entry into the Common Market.

11. The Commonwealth Immigrants Act of 1962 had left no one in doubt as to which part of the Commonwealth (white or black) control was applicable.

12. Already — by 1965 — 40 per cent of all junior hospital medical staff were from the New Commonwealth and nearly 15 per cent of all student nurses. Without that help some hospitals would have had to close just as without Commonwealth immigrants London Transport would be disrupted (See Lord Stonham, Lords *Hansard* 10 March 1965, Col. 96). And David Ennals told the Commons some days later that in 1963 'immigrant teachers, nurses, professional engineers and chemists numbered only half as many as their British counterparts who left for other parts of the world' (*Hansard* 23 March 1965, Col. 393).

13. *Commonwealth Immigrants Act 1962, Control of Immigration Statistics, 1966*. Cmnd. 3258, London: HMSO, 1967.

14. See M. Nikolinakos, 'Uber das Nord-Sud Problem in Europa: das Konzept des Europaischen Sudens', *Dritte Welt*, Heft 1/1974, pp. 29-50.

15. Psychologically it might be — but this is of no interest to capital unless there is profit in it. Socially, it is counter-productive.

16. Jamaica and Trinidad and Tobago achieved independence in August 1962, Guyana in May 1966, Barbados in November 1966. Antigua, Dominica, Grenada, St. Lucia and other small islands became 'non dependent states' in 1967.

17. Roy Hattersley, quoted in *Colour and Citizenship*, *op. cit.*

18. Sheila Patterson, *Immigration and Race Relations in Britain, 1960-67*, London: OUP for IRR, 1969.

19. 'Address given by the Home Secretary, the Rt. Hon. Roy Jenkins, MP, on 23 May 1966 to a meeting of Voluntary Liaison Committees', London, NCCI., 1966.

20. In a speech supporting immigration control, Roy Hattersley MP (Labour) remarked that it was now 'necessary to impose a test which tries to analyse which immigrants, as well having jobs or special skills are most likely to be assimilated into our national life'. This would, he added, favour the English-speaking West Indians as against the Pakistanis (See Patterson, *op. cit.*).

21. *The Times* News Team, *The Black Man in Search of Power*, London: Nelson, 1968.
22. See 'The Liberation of the Black Intellectual', pp. 82-98 above.
23. In 1974 the median settlement was £23.50 (*White Paper on Racial Discrimination*, London, HMSO, September 1975).
24. This refers to the Spaghetti House Siege.
25. 'Gunning for a Wage', *Race Today*, Vol. 7, no. 10, October, 1975.
26. Antonio Gramsci, *Selections from Prison Notebooks*, edited by Quintin Hoare & G. Nowell Smith, London: Lawrence and Wishart, 1971.

Grunwick*

Two recent events have further elucidated the strategies of the state vis-à-vis the black community and, more especially, the black section of the working class, first analysed in 'Race, class and the state'[1] over a year ago. One is the House of Commons Select Committee Report on the West Indian Community and the other is the 10-month-old strike of Asian workers at the Grunwick Film Processing plant in Willesden in North London. Of these, the Grunwick issue is the more complex and confusing and, if only for those reasons, the more challenging of analysis — however risky the exercise of writing history even as it is being made.

Grunwick processes photographic films and relies a great deal on the mail-order business. It is estimated that around 90 per cent of those on the processing side are Asians, many of them women and most of them from East Africa. The strikers first walked out when a worker was sacked after being forced to do a job he could not possibly do in the time alloted for it. This was typical of the punitive, racist and degrading way in which the management treated the workforce. The strikers, on the advice of the local trades council, joined APEX (the Association of Professional Executive Clerical and Computer Staff). The employers, however, refused to recognise the union and the

* *Race & Class*, Vol. 19, no. 1, summer 1977.

strike has now centred on the question of union recognition by management — since union recognition is a prerequisite to raising the wages from the exceptionally low figure of £25 for a 35-hour week.

The strike has received widespread union support, which is in certain respects unique in the history of British trade unionism. Not only has full strike pay from APEX been forthcoming from the very beginning, but also other national unions, e.g. Transport and General Workers Union, the Union of Post Office Workers (UPW), the Trades Union Congress (TUC), and through their encouragement hundreds of local union branches, shop stewards committees, trades councils and others, have given financial and other support. Not only did Len Murray, General Secretary of the TUC, intervene personally in the dispute, but cabinet ministers have themselves been to the picket lines to give their support. After a certain amount of pressure, the UPW took the almost unprecendented step of introducing a postal ban. Although this lasted only four days in the event, it hit management hard since it relies on the mail-order side for 60 per cent of its business.

At first it looked as though Grunwick was to be the rallying point for the labour movement to prove its commitment to black workers. But what is more apparent now is that the unions have been carefully determining the direction that the strike should take and the type of actions open to the strikers. It is worth recalling here the comments of George Bromley, a union negotiator for 30 years with London Transport, who in 1974, during the Imperial Typewriters strike of Asian workers, said, 'The workers have not followed the proper dispute procedures. They have no legitimate grievances and it's difficult to know what they want... Some people must learn how things are done'.

The 'proper procedures' have in this case certainly been taught — and followed to the bureaucratic letter. When the right-wing National Association For Freedom threatened legal action against the postal boycott,[2] the UPW capitulated, arguing to the strikers that they had persuaded management to go to arbitration to the Advisory Conciliation and Arbitration Service (ACAS). But when ACAS called for a ballot of the workforce, management sought to limit it to those still at work and not the strikers — so discrediting the ACAS procedure and the Employment Protection Act within which it operates. Similar bureaucratic procedures, such as appeals to the Industrial Tribunal and recourse to government investigation, have proved equally futile — and, worse, delayed the possibility of effective solidarity action. It was six months before ACAS's report (in favour of the strikers) finally came out. Nor has the UPW reintroduced its ban, despite its promise to do so once the report was out.

On the other hand, the unions have induced the strikers to stay out by almost doubling their strike pay. But while the unions are keen to keep the strike going at all costs, the strikers themselves have begun to question the conduct and purpose of the unions' support. According to Mrs Desai, treasurer of the strike committee, 'If the TUC wanted, this strike could be won tomorrow.' The workers are belatedly resorting to tactics they urged in the first place, such as picketing local chemists shops (from which Grunwick's trade also comes) and organising 24-hour pickets.

Asian workers have over the last two decades proved to be one of the most militant sections of the working class. In strike after strike — Woolf's, Perivale Gütermann, Mansfield Hosiery, Imperial Typewriters, Harwood Cash and others — they have not only taken on the employers and sometimes won (limited) victories, but have also battled against racist trade unions which have either dragged their feet or quite often denied them the support they would have afforded white workers. The Imperial Typewriters case was the most blatant. In May 1974 Asians at Imperial Typewriters (a subsidairy of Litton Industries) went on strike over differentials between white and Asian workers. The unions refused their support and the strikers, supported by other black workers, had to fight both union and management (bolstered by the extreme right-wing party, the National Front).

Over the Grunwick dispute, however, the unions have been unusually supportive of the Asian workforce. Some commentators on the left have traced the union change of direction to a sudden change of heart: it had come upon them (the unions) that racism was a bad thing and should be outlawed from within their ranks. But why this 'change of heart'?

In the first instance, of course, the basis of the Grunwick dispute is the unionisation of the workforce and it is therefore in the interests of the unions (and indeed their business) to recruit workers into their organisations. This is the most obvious reason for union support of the strike. But the inordinate anxiety to unionise the workers must be seen in the larger context of government-trade union collaboration in the Social Contract.

In effect what the government says to the workers in the Social Contract is: 'we are in a time of great economic crisis, with increasing inflation and galloping unemployment. The only way we are going to solve the problem is by keeping wages down. But we can do this only with your agreement to put up with hardships. So if you agree not to use your power of collective action (the only power you really have to improve your conditions) we will in turn see that you are protected from the employers taking advantage of your restraint. We will, in return for your abandoning the right to collective bargaining, give you statutory safeguards to keep the employers at bay.' Hence the Employment Protection Act 1975, the Trade Union and Labour Relations Acts of 1974 and 1976, the Sex Discrimination Act 1975, the Health and Safety at Work Act 1974 and the Equal Pay Act 1970 (enforced in 1975). And, more recently, Michael Foot, Leader of the House of Commons, has inveighed against the judiciary for its apparent anti-union bias. 'If the freedom of the people of this country — and especially the rights of trade unionists — had been left to the good sense and fairmindedness of judges, we would have precious few freedoms in this country.'

The Grunwick dispute, if the other Asians strikes are anything to go by, threatens to blow a hole, however small, in the Social Contract, and in the circumstances (of the rank and file of the working class clearly jibbing at a further extension of the Social Contract), one swallow could easily make a summer! To bring the dispute within the Social Contract framework it is necessary to unionise the Asian strikers. But to unionise a black workforce, it is first necessary to take a stand against racial discrimination. It is necessary to speak to the workers' first and overwhelming 'disability'. 'The strike,' said Mrs Desai, 'is not so much about pay, it is a strike about human dignity.' Hence, if the unions are to win the confidence of the strikers, and of black workers in general, they have to take an unequivocal stand against the employer's racist practices. Besides, it is the very fact of colour that has, as so many times before, lent a political dimension to the struggle of the Grunwick strikers — and the unions, as so many times before, are anxious to keep that dimension out, particularly in view of the Social Contract. Additionally, in the overall strategy of the state, the management of racism in employment has, since the strikes of 1972-74, been handed over to the trades unions (and not to the Community Relations Commission).

Now that the state has decided that the social and political cost of racism has begun — in the objective circumstances — to outweigh its economic profitability (see 'Race, Class and the State'), the unions are equally anxious to contribute to that effort.

In fact as far back as the TUC Conference in September 1976, APEX General Secretary, Roy Grantham, spoke about the Grunwick dispute in the context of the Government White Paper on Racial Discrimination, which heralded the Race Relations Act. The Act itself, passed in November 1976, is very concerned with employment and in fact extends the application of the new employment laws' complaints procedures to the area of racial discrimination. This Act, unlike previous race relations acts, has full union backing.

This support for the new legislation has been accompanied by increased interest and concern about race relations within the trade union bureaucracies since the 1972-4 period of disputes. After the Mansfield Hosiery strike there followed a whole spate of strikes throughout the East Midlands involving Asian workers in dispute not only with management, but usually with the union and fellow white workers too. Strike committees of different factories supported each other, workers were learning from the examples set in neighbouring cities, local black communities supported the strikers, there was serious debate about the need to set up black trades unions. It is since then that we find proposals for special training on shop steward courses, the establishment of race relations departments in national unions and the TUC and the production of a TUC model equal opportunity clause for contracts. And, more recently, a government race relations employment advisory group has been set up on which the TUC and ACAS, as well as the Confederation of British Industry and the Commission for Racial Equality, will be represented.

But the management of racism in employment is not the only thing that has been left to the unions' care. They have also been entrusted with the task of selling the Employment Protection Act to the workforce as a whole. The Grunwick dispute encompasses both these functions.

What we have, therefore, is not a 'change of heart' but a change of tactics — to ordain, legitimise and continue the joint strategies of the state and union leaders against the working class — through the Social Contract.

Notes

1. See pp. 101-26 above.
2. Under the 1953 Post Office Act, which prohibits 'interference with the mail'.

From Immigration Control to 'Induced Repatriation'*

In 'Race, class and the state', we have shown how successive British governments, whether Tory or Labour, have used Nationality Laws and Immigration Acts to adjust the intake of labour into Britain. At one end of the spectrum is the Nationality Act of 1948 which, merely by conferring British citizenship on the citizens of the colonies and the Commonwealth, made sure that post-war Britain laid its hands on all the workers it could get — and if they came as settlers to the mother country, so much surer was Britain of their continued labour. At the other end is the Immigration Act of 1971, which, in response to the economic recession of the 1960s and 1970s, finally terminated all black settler immigration and installed instead a system of con-

* *Race & Class*, vol. XX, no. 1, Summer 1978.

tract labour on the lines of the European *Gastarbeiter*.

With the enforcement of the 1971 Act (in January 1973) all primary immigration virtually ceased. The only immigrants who were allowed in, beyond the specific needs of the British economy, were the dependants of those already settled here and such special categories as UK passport holders from East Africa and 'male fiancés'. It would have appeared therefore that the Dutch auction on immigration had also ceased and that the two major political parties could get down to the more serious business of 'race relations'. For they both held the view that to improve race relations, to make things better for the 'coloureds', you must first restrict their numbers.

But the inexorable logic of that philosophy — which for a while had engendered a bi-partisan policy on immigration and even facilitated the passage of anti-discriminatory legislation in the Race Relations Act of 1976 — finally caught the fancy of Margaret Thatcher in an election year. And at the end of January 1978 she announced that her party would 'finally see an end to immigration' — not just for the sake of race relations (though that too) but in order to preserve the British way of life and allay the majority's fears, for 'this country might be rather swamped by people with a different culture'. Before translating this statement into policy terms, however, the Tories prepared to await the recommendations shortly to be made by the House of Commons (all party) Select Committee on Race Relations and Immigration.

Whether the Select Committee was hustled into Thatcherite recommendations by the Tory woman's speech or had arrived at them spontaneously or through an overwhelming desire for 'consensus' is not certain. What is certain, however, is that the recommendations of a Committee representing all major parties (and 'wings' of parties) have given a 'certificate of respectability' to the Tory proposals that were to follow. In fact when Thatcher's henchman, Willie Whitelaw, came to outline Tory policies on race and immigration some two weeks later, they had the ring of the Select Committee — and the purpose, it would appear, was the same: to move from immigration control to induced repatriation.

The Select Committee Report and Tory proposals

Primary immigration

Scour as it might, the Select Committee could not come up with any fresh ideas for the control of *primary* immigration. A few refinements, though. On the level of Britain's economic needs, a further reduction should be attempted — with regard to doctors, etc. (who at present enter without special permits) and to nursing auxiliaries and hotel and catering workers (mainly southern European). The entry status of husbands and 'male fiancés' should be revised. A new Nationality Law — which is in the offing anyway — would tidy up some of the anomalies and so help to reduce immigration.[1]

The Tories went one better. While pledging themselves to stricter 'work-permit free employment' and 'a new British Nationality Act', they promised to end, without review, the automatic entry of 'non-patrial' husbands and 'non-patrial' male fiancés on the ground that the 'abode of the husband in a marriage should normally be viewed as the natural place of residence of the family'. And as for a quota system, this would not be applied to 'the Indian sub-continent alone in a discriminatory fashion', as suggested by the Select Committee, but to 'all forms of entry from all non-EEC countries right across the board'. Except that, a limited 'register for eligible (sic) wives and children from the Indian subcontinent' would first be established before introducing the 'across the board' quota system!

Secondary immigration

Already in these recommendations and proposals the Select Committee and the Tory party are beginning to slide over from 'primary immigration' into the immigration of dependants. And it is here in the matter of dependants that the axe really falls. The report recommends that children over 12 years old born abroad to those settled here should not be allowed to join their parents (para. 143), and 'it may well be necessary on social grounds to adjust the Immigration Rules in the future to ensure children are only admitted if they are below school age' (142). Dependants such as parents, grandparents and children over 18 should not be allowed to join families unless accommodation and means of support by the sponsor is approved by the authorities.

Again, the Tories do better — except in regard to the recommendation about children over 12 being refused entry. They pledge to 'restrict the entry of parents, grandparents and children over 18 to those who can prove urgent compassionate grounds'. And as for those who entered Britain after 1 January 1973, they would not only forego automatic right of settlement but would, even if allowed to settle, have 'no automatic entitlement for wives and dependants'.

The message is clearly that unproductive additions to working-class black families are unwanted. If you want family life, 'go home'.

From surveillance to pass laws

But perhaps the most damaging aspect of the report is the recommendations it makes for the surveillance of the black community. Already, from the village in the Indian subcontinent to the British social security office, blacks are checked and scrutinised, women sexually examined. The Committee now recommends that the Department of Health and Social Security should 'introduce without delay new procedures to tighten up identity checks', (87) 'the police, the Immigrants Service Intelligence Unit and other authorities' should be given powers to seek out illegal immigrants and overstayers, (86) and new powers should be sought 'to provide effective sanctions' against employers who knowingly employ them. And finally the government is recommended to hold an independent inquiry to consider 'a system of internal control of immigration'.(89)

The Tories for their part 'agree' with the Select Committee's recommendation for the setting up of an inquiry into such a 'system of internal control' — a system presumably of identity cards and pass laws for those legally here — but promise in the meantime to 'intensify counter measures against illegal immigration and overstaying'. They will do even more: they will 'improve' existing 'arrangements... to help those who are really anxious to leave this country' — voluntary repatriation of course, but aided.

Again the message to the black community is clear: if you

want to live in peace, go home.

Quite clearly all the talk about primary and secondary immigration is a lot of nonsense. What the Select Committee and the Tory party really want is to reduce the number of 'coloured immigrants', irrespective of whether they are dependants or not. And since primary immigration, as the Labour government has pointed out, is 'down to a trickle' — new workers on permits in 1976 were something in the order of 0.5 per cent of the total immigration — the cuts have to be made on those overseas, mainly from the Indian subcontinent,[2] waiting to be reunited with their families here. The result, whatever the aim, is to divide black families even further — leaving it to the wage-earner to face the impossible question whether he wants to go on working and living in Britain if he cannot have his dependants with him — those for whom he came to work here in the first place. But even if, for the sake of their livelihood and his, he is forced to remain, he would be subjected to 'a system of internal control' (for how do you distinguish between an 'illegal immigrant' and a legal one when they are all black) that would make life untenable.

The new philosophy

The crude economics of the strategy would indicate that Britain no longer needs the workforce it was once desperate for and would like to lay them off — preferably in their countries of origin. All the Immigration Acts from 1962 onwards had pointed to this end — culminating in the Act of 1971 which finally established a system of importing workers when they were needed and sending them back when they were not. And all the Immigration Acts — because they were aimed at 'coloured immigrants' and therefore tied to race — were preceded by and sought justification in the philosophy that fewer blacks make for better race relations. The economic strategy, in other words, threw up the philosophy and the philosophy engendered the Acts. But the 1971 Act applied to future immigration; there was always the question of what to do with those who had settled here before then. So that even when primary immigration was virtually ended, the philosophy continued — to be applied this time to the dependants of those who had settled here.[3] But alongside that, another philosophy, springing from the economics of recession — still based on the premise that fewer blacks make for better race relations — begins to emerge: one that would eventually provide the justification for repatriation.

In the deliberations of the Select Committee, the new philosophy is only half formed, and tempered with reason. Minorities, they say, should pay 'greater regard to the *mores* of their country of adoption' — if only because such customs as arranged marriages encourage the immigration of 'male fiancés. Of course, if the immigrants wished to maintain such a custom, they should be encouraged to observe their own 'traditional pattern' of the bride joining the husband's family (160). And in that same spirit of sparing immigrants any further culture shock the Committee recommends that, in future, children should only be admitted if they are below school age (142). But even so, all that the Select Committee seems to be saying is: if you are here, be like us, if you cannot, go home.

In the hands of the Tories, however, the new philosophy (coupled with the old) emerges full fledged and stark:[4] people of an alien culture are threatening to 'swamp' Britain, the British way of life is in danger, the majority's fears must be allayed. Hence the Tory proposals. But lest they did not go far enough, the minion Whitelaw declared in effect that there could not be one law for the white and another for the black — and policies of 'reverse discrimination' which put minorities in a privileged position should be rejected because of 'the dangers to racial harmony'. Let them learn English instead — through language classes, in factories and workplaces — and this will help them to get employment and promotion. And in the same vein Mrs Thatcher has pledged to amend the 'incitement to racial hatred' clause in the Race Relations Act (1976) so that 'intent to offend' has to be proved to form the basis of a prosecution.[5] Indeed she will 'look at the whole Act when we come into power'.

Repercussions

But if Tory pledges still appear to fall short of the emerging Tory philosophy, it is because in British race policy, the philosophy, the rationale, comes first — then to be taken up and embellished by sections of the media — thereby creating a 'climate of opinion' — demanding legislation.[6] In the days following Thatcher's interview (30 January) the *Daily Mail*, in a series of articles on immigration with titles such as 'They've taken over my home town', gave 'real life stories' of culture swamping. Four days before the Select Committee report (13 March) BBC television, in a programme meant to elicit a whole range of views on immigration, afforded Enoch Powell a field day during which he enlarged on his idea of 'induced repatriation'. The Tory proposals were published on 7 April and in the countrywide local elections (4 May) Tory candidates were able to reiterate and justify Tory proposals.

All of which has gone to create a climate of opinion favourable to Tory immigration policies and Tory philosophy. But if legislation, at central government level, has still to await the results of a general election, no such impediment stands in the way of local governments with a Tory majority. One such authority, Slough, has offered a homeless woman (married to 'an American serviceman') and her child a loan to leave the country rather than meet its obligation to house her.[7] The deputy leader of another Tory-controlled council has warned its community workers that publishing advice to squatters or advising West Indians (and other groups) on what to do when arrested for suspicious loitering [8] or 'demonstrating with the Grunwick pickets' could constitute political activity, incurring cuts in staff and finance.

But the most devastating consequence of all this Tory activity and, more particularly (because more appealing), the shadow Prime Minister's lurid statement about culture swamping, has been the terror unleashed on the black communities. Down in the ghetto — far removed from Thatcher country — the majority, whose fears the Tory lady wished to allay, have literally taken the law into their own hands and begun to terrorise the black population. And it has not stopped at knifings, beatings and harrassments. Altab Ali, a Bengali going home from work, was stabbed to death in the East End of London. On his tenth birthday, Kennith Singh, returning from the corner shop with his mother's cigarettes, was battered to death and dumped on a rubbish tip in Plaistow. And shot gun attacks have been mounted on West Indians in Wolverhampton.

For the white maggots that feed on the psychopathic fancies of a super race, 'go home, nigger' is no longer a dream but a promise which, to be fulfilled, needs only their unremitting activity.

Politics and ideology

Viewed in terms of electoral politics, the Tory tactics would appear to be no more than a vote catching exercise, cynically exploiting the racialism of the white working class. But the philosophy from which the tactics emanate is ingrained in the philosophy of archaic, free-enterprise capitalism, which Margaret Thatcher and her cohorts espouse. For enlightened capital, the appeal to narrow nationalism and racialism is not only unnecessary but, in terms of social cost, counter-productive. Hence the Race Relations Act of 1976 was aimed at dismantling institutional racism (its economic function was over) while still allowing racialism to divide the working class. Since then, however, the Labour government in its role as agent of the ruling class, has persuaded the unions, the alleged representatives of the working class, to adopt its interests as their own in a social contract. And to leave the black working class out of such a contract would be to invite a militancy which could infect the rest of the class. (This was precisely the significance of the Grunwick strike.)[9] At the same time the rise of the National Front has shown that the continuing division of the working class on racialist lines not only weakens Labour's electoral base but gives also a fillip to fascism — which is not in the interests of corporate capital: you don't need to defeat the working class when you can co-opt it. Hence Labour has disassociated itself from the recommendations of the Select Committee, inveighed against Tory immigration proposals as unnecessary, 'inhumane' and pointing to a 'pass law society', and been visibly active in ongoing anti-racist, anti-fascist campaigns.

Tory thinking, however, has continued to remain in the past.[10] The racist and anti-working class sentiments of Thatcher and Joseph (he who inveighed against lower class fertility rates) reflect the anxieties and fears of a decaying bourgeoisie. And not unnaturally they figure that to defeat the class by dividing it there is no better ploy than race — which while speaking to the social and economic frustrations of white workers, provides them at the same time with a national vision, the surrogate for revolution. The strategy is to defeat the class, the tactic is to divide it — and they both stem from the ideology of outmoded capital.

In the process, however, the Tories have stepped into National Front country — in National Front guise. But given their ideology, they cannot (to mix a metaphor) deliver the NF goods. Their thinking on race, for instance — even in its 'induced repatriation' phase — cannot proceed to its logical conclusion of compulsory repatriation or to the final solution provided by fascism. (How else would you stop the swamping of British culture by the blacks?) What it does, however, is to lend orthodoxy to fascist thinking and soften up the working class for fascist blandishments — and fascist ideology.

Notes

1. This became law with the Nationality Act of 1981.
2. More West Indians are leaving Britain than coming in.
3. If the proposed register of dependants to establish a quota system is anything to go by, the current concern to ascertain immigrant birthrate, and hence the size of the 'coloured population' at the end of the century, is an ominous sign. That children born here would be British at least by the year 2000, however, has escaped statistical consideration.
4. That the statement of Tory philosophy was made (by Margaret Thatcher) before the Select Committee report is not to the point: one did not grow out of the other, but they were both travelling in the same direction at the speed of their respective motivations. What is to the point, however, is that the Select Committee report justified the Tory statement that preceded it and gave a fillip to the Tory proposals that followed it.
5. The prosecution of Kingsley Read for his notorious statement on the murder of a young Asian — 'one down, a million to go' — foundered precisely because, under the 1965 Act then in operation, *intent* to inflame racial hatred had to be proved.
6. See the 'Editorial' and 'UK commentary' in *Race & Class* Vol. XVIII, no. 1, 1976.
7. 'For want of a better word this is a repatriation scheme... It's easier

and cheaper to pay their fares than give them houses at the expense of Slough people,' R. Stephenson, deputy leader, Slough Council (*Glasgow Herald*, 29 April 1978).

8. The notorious 'Sus' law is one of the principal weapons used by the police to harass young blacks. (Almost half the arrests for 1976 in the London area were of blacks.)

9. See 'Grunwick', pp. 126-31 above, and *Race & Class*, Vol. XIX, no. 3, 1978.

10. Of course, the Tory party, like any other, is not homogeneous — and contains within it the more long-term conservatism of Peter Walker, Ted Heath et al.

Part Four

Racism and Imperialism

These two essays bring together analysis of racism and imperialism. Developing from the earlier analysis of labour migration and the function of racism, 'Imperialism and Disorganic Development in the Silicon Age' examines the significance of a fundamental shift in the nature of imperialism. From a system in which labour moved to the centres of capital, imperialism had developed into a system in which capital moves to wherever there is an accessible source of cheap labour — a process which has been expanded and accelerated by the revolutionary technology of the microchip. While other studies have looked at one element or another — examined the international division of labour, or attempted to understand the implications of the new technology for employment, or looked at labour migration or the operation of multinationals and their client regimes in the Third World — nowhere else has the whole range of the change been examined (albeit briefly) for its political, cultural and economic implications.

'Race, Caste and Class in South Africa' is a historical examination of racism in different social formations. It analyses the interrelationship of these key concepts in a way that is of importance not only for theoretical analysis, but for its implications for political struggle. The debate on the national question is crucial not only in respect of South Africa, but for radical analysis and practice in the Third World.

Imperialism and Disorganic Development in the Silicon Age*[1]

One epoch does not lead tidily into another. Each epoch carries with it a burden of the past — an idea perhaps, a set of values, even bits and pieces of an outmoted economic and political system. And the longer and more durable the previous epoch the more halting is the emergence of the new.

The classic centre-periphery relationship as represented by British colonialism — and the inter-imperialist rivalries of that period — had come to an end with the second world war. A new colonialism was emerging with its centre of gravity in the United States of America; a new economic order was being fashioned at Bretton Woods. Capital, labour, trade were to be unshackled of their past inhibitions — and the world opened up to accumulation on a scale more massive than ever before. The instruments of that expansion — the General Agreement on Tariffs and Trade, the International Monetary Fund and the World Bank — were ready to go into operation.[2] Even so, it took the capitalist nations of western Europe, Japan and the United States some twenty-five years to rid themselves of the old notions of national boundaries and 'lift the siege against multinatinal enterprises so that they might be permitted to get on with the unfinished business of developing the world economy' (Rockefeller). The Trilateral Commission was its acknowledgement.

Britain, hung up in its colonial past, was to lag further behind. It continued, long after the war, to seek fresh profit from an old relationship — most notably through the continued exploitation of colonial labour, but this time at the centre. So that when the rest of Europe, particularly Germany, was reconstructing its industries and infrastructure with a judicious mix of capital and labour (importing labour as and when required), Britain, with easy access to cheap black labour and easy profit from racial exploitation, resorted to labour-intensive production. And it was in the nature of that colonial relationship that the immigrants should have come as settlers and not as

---

* *Race & Class*, vol. XXIV, no. 2, Autumn 1979.

labourers on contract.

The history of British immigration legislation including the present calls for repatriation is the history of Britain's attempt to reverse the colonial trend and to catch up with Europe and the new world order.[3]

That order, having gone through a number of overlapping phases since the war, now begins to emerge with distinctive features. These, on the one hand, reflect changes in the international division of labour and of production, involving the movement of capital to labour (from centre to periphery) which in turn involves the movement of labour as between the differing peripheries. On the other hand, they foreshadow a new industrial revolution based on micro-electronics — and a new imperialism, accelerating the 'disorganic' development of the periphery. And it is to these new developments in capitalist imperialism that I want to address myself, moving between centre and periphery — and between peripheries — as the investigation takes me, bearing in mind that these are merely notes for further study.

The early post-war phase of this development need not detain us here, except to note that the industrialisation undertaken by the newly independent countries of Asia and Africa (Latin America had begun to industrialise between the wars) put them further in hock to foreign capital, impoverished their agriculture and gave rise to a new bourgeoisie and a bureaucratic elite.[4] The name of the game was import substitution, its end the favourable balance of trade, its economic expression state capitalism, its political raison d'être bourgeois nationalism. Not fortuitously, this period coincided with the export of labour to the centre.

Capital and labour migration

By the 1960s, however, the tendency of labour to move to capital was beginning to be reversed. The post-war reconstruction of Europe was over, manufacturing industries showed declining profit margins and capital was looking outside for expansion. The increasing subordination of Third World economies to multinational corporations made accessible a cheap and plentiful supply of labour in the periphery, in Asia in particular. Advances in technology — in transport, communications, information and data processing and organisation — rendered geographical distances irrelevant and made possible the movement of plant to labour, while ensuring centralised control of production. More importantly, technological development had further fragmented the labour process, so that the most unskilled worker could now perform the most complex operations.

For its part, the periphery, having failed to take off into independent and self sustained growth through import substitution,[5] turned to embrace export-oriented industrialisation — the manufacture of textiles, transistors, leather goods, household appliances and numerous consumer items. But capital had first to be assured that it could avail itself of tax incentives, repatriate its profits, obtain low-priced factory sites and, not least, be provided with a labour force that was as docile and undemanding as it was cheap and plentiful. Authoritarian regimes, often set up by American intervention, provided those assurances — the Free Trade Zones provided their viability.[6]

The pattern of imperialist exploitation was changing — and with it, the international division of production and of labour. The centre no longer supplied the manufactured goods and the periphery the raw materials. Instead the former provided the plant and the know-how while the latter supplied primary products and manufactures. Or, as the Japanese Ministry of Trade in its 'Long Term Vision of Industrial Structure' expressed it, Japan would retain 'high-technology and knowledge-intensive industries' which yielded 'high added value' while industries 'such as textiles which involve a low degree of processing and generate low added value [would] be moved to developing countries where labour costs are low'. As Samir Amin put it in *Imperialism and Unequal Development*, 'the centre of gravity of the exploitation of labour by capital (and in the first place, by monopoly capital which dominates the system as a whole) had been displaced from the centre of the system to the periphery'.

The parameters of that new economic order are best expressed in the purpose and philosophy of the Trilateral Commission. Founded in 1973, under the sponsorship of David Rockefeller of the Chase Manhattan Bank, the Commission brought together representatives of the world's most powerful banks, corporations, communications conglomerates, and international organisations plus top politicians and a few 'free' trade unions and trade union federations (from North America, Europe and Japan) to reconcile the contradictions of transnational capital, while at the same time checking 'the efforts of national governments to seize for their own countries a disproportionate share of the benefits generated by foreign direct investment'.[7] As Richard Falk puts it: 'The vistas of the Trilateral Commission can be understood as the ideological perspective representing the transnational outlook of the multinational corporation' which 'seeks to subordinate territorial politics to non-territorial economic goals'.[8]

And for the purpose of that subordination, it was necessary to distinguish between the differing peripheries: the oil-producing countries and the 'newly-industrialising' countries, and the underdeveloped countries proper (which the Commission terms the 'Fourth World').

The implications of this new imperial ordinance for labour migration — not, as before, between centre and periphery but as between the peripheries themselves — are profound, the consequences for these countries devastating. The oil-rich Gulf states, for instance, have sucked in whole sections of the working population, skilled and semi-skilled, of South Asia, leaving vast holes in the labour structure of these countries. Moratuwa, a coastal town in Sri Lanka, once boasted some of the finest carpenters in the world. Today there are none — they are all in Kuwait or in Muscat or Abu Dhabi. And there are no welders, masons, electricians, plumbers, mechanics — all gone. And the doctors, teachers, engineers — they have been long gone — in the first wave of post-war migration to Britain, Canada, USA, Australia, in the second to Nigeria, Zambia, Ghana. Today Sri Lanka, which had the first free health service in the Third World and some of the finest physicians and surgeons, imports its doctors from Marcos' Philippines. What that must do to the Filipino people is another matter, but all that we are left with in Sri Lanka is a plentiful supply of unemployed labour, which is now being herded into the colony within the neo-colony, the Free Trade Zone.

Or take the case of Pakistan, which shows a similar pattern of emigration, except that being a Muslim country the pull of the

Gulf is even stronger. Besides, the export of manpower — as a foreign exchange earner — is official policy, a Bureau of Emigration having been set up in 1969 to facilitate employment overseas. Consequently Pakistan 'is being progressively converted into a factory producing skilled manpower for its rich neighbours'.[9]

But the export of skilled workers is not the only drain on Pakistan's resources. Apart from its traditional export of primary products, its physical proximity to the oil-rich countries has meant also the smuggling out of fresh vegetables, the sale of fish in mid-seas and the export, often illegal, of beef and goat meat. (The Gulf states raise no cattle.) 'The adverse effects of this trade,' laments Feroz Ahmed, 'can be judged from the fact that Pakistan has one of the lowest per capita daily consumptions of animal protein in the world: less than 10 grammes.'[10]

The Middle East countries in turn have only invested in those enterprises which are geared to their own needs (textiles, cement, fertiliser, livestock) and rendered Pakistan's economy subservient to their interests. And to make this 'development of underdevelopment' palatable they harked back to a common culture. Iranian cultural centres sprouted in every major town in Pakistan, outdoing the Americans, and the teaching of Arabic and Persian was fostered by official policy. 'We the Pakistanis and our brethren living in Iran,' wrote a Pakistani paper, 'are the two Asiatic branches of the Aryan Tree who originally lived in a common country, spoke the same language, followed the same religion, worshipped the same gods and observed the same rites... Culturally we were and are a single people.'[11]

But if Pakistan has been relegated, in the pecking order of imperialism, 'to the status of a slave substratum upon which the imperialist master and their privileged clients play out their game of plunder and oppression',[12] the privileged clients themselves exhibit a distorted 'development'. Take Kuwait for instance. In the pre-oil era Kuwait's economy was based on fishing, pearling, pasturing, trade and a little agriculture. Today all these activities, with the exception of fishing, have virtually ceased — and fishing has been taken over by a company run by the ruling family. The oil industry, while providing the government with 99 per cent of its income, affords employment only to a few thousand. Almost three-fourths of the native work force is in the service sector, with little or nothing to do. (A UN survey estimated that the Kuwaiti civil servant works 17 minutes a day.)[13] But more than 70 per cent of the total work force and over half the total population consists of non-Kuwaiti immigrant labour. And they are subjected to harsh conditions of work, low wages, no trade union rights, wretched housing and arbitrary deportation. Kuwait is, in effect, two societies, but even within the first 'the ruling elite lives in a swamp of consumer commodities and luxuries, while those at the bottom of the Kuwaiti social pyramid are being uprooted from their traditional productive activities and thrown on the market of unproductiveness'.[14]

The pattern of labour migration in South-east Asia is a variation on the same imperial theme, and its consequences no less devastating. The first countries to industrialise in this region were Taiwan in the 1950s and, in the 1960s, South Korea, Singapore and Hong Kong. Taiwan and South Korea were basically offshore operations of the USA and Japan — and, by virtue of their strategic importance to America, were able to develop heavy industry (shipbuilding, steel, vehicles) and chemicals in addition to the usual manufacture of textiles, shoes, electrical goods, etc. And by the middle of the 1970s, these two countries had gone over from being producers of primary products to producers of manufactured goods. Singapore's industrialisation includes ship repair (Singapore is the fourth largest port in the world) and the construction industry. Hong Kong, the closest thing to a 'free economy', is shaped by the world market.

What all these countries could offer multinational capital, apart from a 'favourable climate of investment' (repatriation of profit, tax holidays, etc), was authoritarian regimes with a tough line on dissidence in the work force and a basic infrastructure of power and communications. What they did not have was a great pool of unemployed workers. That was provided by the neighbouring countries.

Hong Kong uses all the migrant labour available in the region, including workers from mainland China, and is currently negotiating with the Philippines government for the import of

Filipino labour. South Korea's shortage of labour, by the very nature of its development, has been in the area of skilled workers. (Not illogically South Korea has been priced out of its own skilled workers, some 70,000 of them, by the developing oil-rich countries of the Middle East.) But it is Singapore which is the major employer of contract labour — from Malaysia mostly (40 per cent of the industrial work force) and also from Indonesia, the Philippines and Thailand — and that under the most horrendous conditions. For apart from the usual strictures on *Gastarbeiters* that we are familiar with in Europe, such as no right of settlement, no right to change jobs without permission and deportation if jobless, Singapore also forbids these workers to marry, except after five years, on the showing of a 'clean record', and then with the permission of the government — and that on signing a bond that both partners will agree to be sterilised after the second child is born. Lee Kuan Yew, with a nod to Hitler, justifies the policy on the ground that 'a multiple replacement rate right at the bottom' leads to 'a gradual lowering of the general quality of the population'.[15] Their working conditions too are insanitary and dangerous and makeshift shacks on worksites (like the bidonvilles) provide their only housing.

And yet the plight of the indigenous workers of these countries is not much better. The economic miracle is not for them. Their lives contrast glaringly with the luxury apartments, automobiles and swinging discos of the rich. To buy a coffee and sandwich on a thoroughfare of Singapore costs a day's wage, in South Korea 12- and 13-year-old girls work 18 hours a day, 7 days a week, for £12 a month, and Hong Kong is notorious for its exploitation of child labour.[16]

How long the repressive regimes of these countries can hold down their work force on behalf of international capital is a moot point — but multinationals do not wait to find out. They do not stay in one place. They gather their surplus while they may and move on to new pastures their miracles to perform.

The candidates for the new expropriation were Indonesia, Thailand, Malaysia and the Philippines whose economies were primarily based on agriculture and on extractive industries such as mining and timber. Like the first group of countries they too could boast of authoritarian regimes — ordained by the White

House, fashioned by the Pentagon and installed by the CIA — which could pave the way for international capital. Additionally, they were able to provide the cheap indigenous labour which the other group had lacked — and the Free Trade Zones to go with it. What they did not have, though, was a developed infrastructure.

Multinationals had already moved into these countries by the 1970s and some industrialisation was already under way. What accelerated that movement, however, was the tilt to cheap labour, as against a developed infrastructure, brought about by revolutionary changes in the production process.

To that revolution, variously described as the new industrial revolution, the third industrial revolution and the post-industrial age, I now turn — not so much to look at labour migration as labour polarisation — between the periphery and the centre, and within the centre itself, and its social and political implications in both.

Capital and labour in the silicon age

What has caused the new industrial revolution and brought about a qualitative leap in the level of the productive forces is the silicon chip or, more accurately, the computer-on-a-chip, known as the microprocessor. (You have already seen them at work in your digital watch and your pocket calculator.)

The ancestry of the microprocessor need not concern us here, except to note that it derives from the electronic transistor, invented by American scientists in 1947 — which in turn led to the semi-conductor industry in 1952-53 and in 1963, to the integrated circuit industry. Integrated circuits meant that various electronic elements such as transistors, resistors, diodes, etc. could all be combined on the tiny chip of semi-conductor silicon, 'which in the form of sand is the world's most common element next to oxygen'.[17] But if industrially the new technology has been in existence for sixteen years, it is only in the last five that it has really taken off. The periodisation of its development is important because it is not unconnected with the postwar changes in the international division of production and of labour and the corresponding movements and operations of the multinational corporations.

The microprocessor is to the new industrial revolution what

steam and electricity was to the old — except that where steam and electric power replaced human muscle, microelectronics replaces the brain. That, quite simply, is the measure of its achievement. Consequently, there is virtually no field in manufacturing, the utilities, the service industries or commerce that is not affected by the new technology. Microprocessors are already in use in the control of power stations, textile mills, telephone-switching systems, office-heating and typesetting as well as in repetitive and mechanical tasks such as spraying, welding, etc. in the car industry. Fiat, for instance, has a television commercial which boasts that its cars are 'designed by computers, silenced by lasers and hand-built by robots' — to the strains of Figaro's aria (from Rossini). Volkswagen designs and sells its own robots for spot welding and handling body panels between presses. Robots, besides, can be re-programmed for different tasks more easily than personnel can be re-trained. And because microprocessors can be re-programmed, automated assembly techniques could be introduced into areas hitherto immune to automation, such as batch production (which incidentally constitutes 70 per cent of the production in British manufacturing). From this has grown the idea of linking together a group of machines to form an unmanned manufacturing system, which could produce anything from diesel engines to machine tools and even aeroengines. And 'once the design of the unmanned factory has been standardised, entire factories could be produced on a production line based on a standard design'.[18] The Japanese are close to achieving the 'universal factory'.

A few examples from other areas of life will give you some idea of the pervasiveness of microelectronics. In the retail trade, for instance, the electronic cash register, in addition to performing its normal chores, monitors the stock level by keeping tabs on what has been sold at all the terminals and relays that information to computers in the warehouse which then automatically move the necessary stocks to the shop. A further line-up between computerised check-outs at stores and computerised bank accounts will soon do away with cash transactions, directly debiting the customer's account and crediting the store's. Other refinements such as keeping a check on the speed and efficiency of employees have also grown out of such computerisation — in

Denmark, for instance (but it has been resisted by the workers).

There are chips in everything you buy — cookers, washing machines, toasters, vacuum cleaners, clocks, toys, sewing machines, motor vehicles — replacing standard parts and facilitating repair: you take out one chip and put in another. One silicon chip in an electronic sewing machine for example replaces 350 standard parts.

But it is in the service sector, particularly in the matter of producing, handling, storing and transmitting information, that silicon technology has had its greatest impact. Up to now automation has not seriously affected office work which, while accounting for 75 per cent of the costs in this sector (and about half the operating costs of corporations), is also the least productive, thereby depressing the overall rate of productivity. One of the chief reasons for this is that office work is divided into several tasks (typing, filing, processing, retrieving, transmitting and so forth) which are really inter-connected. The new technology not only automates these tasks but integrates them. For example, the word processor, consisting of a keyboard, a visual display unit, a storage memory unit and a print-out, enables one typist to do the work of four while at the same time reducing the skill she needs. Different visual display units (VDUs) can then be linked to the company's mainframe computer, to other computers within the company (via computer network systems) and even to those in other countries through satellite communication — all of which makes possible the electronic mail and the electronic funds transfer (EFT) which would dispense with cash completely.

What this link-up between the office, the computer and telecommunications means is the 'convergence' of previously separate industries. 'Convergence' is defined by the Butler Cox Foundation as 'the process by which these three industries are coming to depend on a single technology. They are becoming, to all intents and purposes, three branches of a single industry'.[19] But 'convergence' to you and me spells the convergence of corporations, horizontal (and vertical) integration, monopoly. A 'convergence' of Bell Telephones and IBM computers would take over the world's communication facilities. (Whether the anti-trust laws in America have already been bent to enable such a development I do not know, but it is only a matter of time.)

Underscoring the attributes and applications of the microprocessor is the speed of its advance and the continuing reduction in its costs. Sir Ieuan Maddock, Secretary of the British Association for the Advancement of Science, estimates that 'in terms of the gates it can contain, the performance of a single chip has increased ten thousandfold in a period of 15 years'. And of its falling cost, he says, 'the price of each unit of performance has reduced one hundred thousandfold since the early 1960s'.[20]

'These are not just marginal effects,' continues Sir Ieuan, 'to be absorbed in a few per cent change in the economic indicators — they are deep and widespread and collectively signal a fundamental and irreversible change in the way the industrialised societies will live... Changes of such magnitude and speed have never been experienced before.'[21]

The scope of these changes have been dealt with in the growing literature on the subject.[22] But they have mostly been concerned with the prospects of increasing and permanent unemployment, particularly in the service industries and in the field of unskilled manual employment — in both of which blacks and women predominate.[23] A study by Siemens estimates that 40 per cent of all office work in Germany is suitable for automation — which, viewed from the other side, means a 40 per cent lay-off of office workers in the next ten years. The Nora report warns that French banking and insurance industries, which are particularly labour intensive, will lose 30 per cent of their work force by 1990. Unemployment in Britain is expected to rise by about 3 million in that time.[24] Other writers have pointed to a polarisation in the work force itself — as between a small technological elite on the one hand and a large number of unskilled, unemployable workers, counting among their number those whose craft has become outmoded. Or, as the Chairman of the British Oil Corporation, Lord Kearton, puts it: 'we have an elite now of a very special kind at the top on which most of mankind depends for its future development and the rest of us are more or less taken along in the direct stream of these elite personnel'.[25]

All the remedies that the British Trades Union Congress has

been able to suggest are 'new technology agreements' between government and union, 'continuing payments to redundant workers related to their past earnings' and 'opportunities for linking technological change with a reduction in the working week, working year and working life time'.[26] The Association of Scientific Technical and Managerial Staffs (ASTMS), whose members are more immediately affected by automation, elevates these remedies into a philosophy which encompasses a changed attitude towards work that would 'promote a better balance between working life and personal life', 'recurrent education throughout adult life' and a new system of income distribution which in effect will 'pay people not to work'.[27]

But, in the performance, these are precisely the palliatives that enlightened capitalism (i.e. multinational capitalism as opposed to the archaic private enterprise capitalism of Margaret Thatcher and her mercantile minions) offers the working class in the silicon age. Translated into the system's terms, 'new technology agreements' mean a continuing social contract between the unions and the government wherein the workers abjure their only power, collective bargaining (and thereby take the politics out of the struggle) and a new culture which divorces work from income (under the guise of life-long education, part-time work, early retirement, etc.) and provides the raison d'être for unemployment. Already the protagonists of the establishment have declared that the Protestant work ethic is outdated (what has work got to do with income?), that leisure should become a major occupation (university departments are already investigating its 'potential'), that schooling is not for now but for ever.

I am not arguing here against technology or a life of creative leisure. Anything that improves the lot of mankind is to be welcomed. But in capitalist society such improvement redounds to the few at cost to the many. That cost has been heavy for the working class in the centre and heavier for the masses in the periphery. What the new industrial revolution predicates is the further degradation of work where, as Braverman so brilliantly predicted, thought itself is eliminated from the labour process,[28] the centralised ownership of the means of production, a culture of reified leisure to mediate discontent and a political system incorporating the state, the multinationals, the trade unions, the bureaucracy and the media, backed by the forces of 'law and order' with microelectronic surveillance at their command. For in as much as liberal democracy was the political expression of the old industrial revolution, the corporate state is the necessary expression of the new. The qualitative leap in the productive forces, ensnared in capitalist economics, demands such an expression. Or, to put it differently, the contradiction between the heightened centralisation in the ownership of the means of production — made possible not only by the enormous increase in the level of productivity but also by the technological nature of that increase — and the social nature of production (however attenuated) can no longer be mediated by liberal democracy but by corporatism, with an accompanying corporate culture, and state surveillance to go with it.

But nowhere is there in the British literature with the exception of the CIS report,[29] any hint of a suggestion that the new industrial revolution, like the old, has taken off on the backs of the workers in the peripheries — that it is they who will provide the 'living dole' for the unemployed of the West. For, the chip, produced in the pleasant environs of 'Silicon Valley' in California, has its circuitry assembly in the toxic factories of Asia. Or, as a Conservative Political Centre publication puts it, 'while the manufacture of the chips requires expensive equipment in a dust-free, air-conditioned environment, little capital is necessary to assembly them profitably into saleable devices. And it is the assembly that creates both the wealth and the jobs.'[30]

Initially the industry went to Mexico, but Asia was soon considered the cheaper. (Besides 'Santa Clara was only a telex away'.) And even within Asia the moves were to cheaper and cheaper areas: from Hong Kong, Taiwan, South Korea and Singapore in the 1960s, to Malaysia in 1972, Thailand in 1973, the Philippines and Indonesia in 1974 and soon to Sri Lanka. 'The manager of a plant in Malaysia explained how profitable these moves had been: "one worker working one hour produces enough to pay the wages of 10 workers working one shift plus all the costs of materials and transport".'[31]

But the moves the industry makes are not just from country to country but from one batch of workers to another within the country itself. For, the nature of the work — the bonding under a microscope of tiny hair-thin wires to circuit boards on wafers of silicon chip half the size of a fingernail — shortens working life. 'After 3 or 4 years of peering through a microscope,' reports Richael Grossman, 'a worker's vision begins to blur so that she can no longer meet the production quota.'[32] But if the microscope does not get her ('grandma where are your glasses' is how electronic workers over 25 are greeted in Hong Kong), the bonding chemicals do.[33] And why 'her'? Because they are invariably women. For, as a Malaysian brochure has it, 'the manual dexterity of the oriental female is famous the world over. Her hands are small and she works fast with extreme care. Who, therefore, could be better qualified by nature and inheritance to contribute to the efficiency of a bench assembly production line than the oreintal girl?'[34]

To make such intense exploitation palatable, however, the multinationals offer the women a global culture — beauty contests, fashion shows, cosmetic displays and disco dancing — which in turn enhances the market for consumer goods and western beauty products. Tourism reinforces the culture and reinforces prostitution (with package sex tours for Japanese businessmen), drug selling, child labour. For the woman thrown out of work on the assembly line at an early age, the wage earner for the whole extended family, prostitution is often the only form of livelihood left.[35]

A global culture then, to go with a global economy, serviced by a global office the size of a walkie-talkie held in your hand[36] — a global assembly line run by global corporations that move from one pool of labour to another, discarding them when done — high technology in the centre, low technology in the peripheries — and a polarisation of the workforce within the centre itself (as between the highly skilled and unskilled or deskilled) and as between the centre and the peripheries, with qualitatively different rates of exploitation that allow the one to feed off the other — a corporate state maintained by surveillance for the developed countries, authoritarian regimes and gun law for the developing. That is the size of the new world order.

Disorganic development

But it is not without its contradictions. Where those contradictions are sharpest, however, are where they exist in the raw — in the peripheries.[37] For what capitalist development has meant to the masses of these countries is increased poverty, the corruption of their cultures, and repressive regimes. All the GNP they amass for their country through their incessant labour leaves them poorer than before. They produce what is of no real use to them and yet cannot buy what they produce — neither use value nor exchange value — neither the old system nor the new.

And how they produce has no relation to how they used to produce. They have not grown into the one from the other. They have not emerged into capitalist production but been flung into it — into technologies and labour processes that reify them and into social relations that violate their customs and their codes. They work in the factories, in town, to support their families, their extended families, in the village — to contribute to the building of the village temple, to help get a teacher for the school, to sink a well. But the way of their working socialises them into individualism, nuclear families, consumer priorities, artefacts of capitalist culture. They are caught between two modes, two sets, of social relations, characterised by exchange value in the one and use value in the other — and the contradiction disorients them and removes them from the centre of their being. And not just the workers, but the peasants too have not escaped the capitalist mode. What it has done is to wrench them from their social relations and their relationship with the land. Within a single life-time, they have had to exchange sons for tractors and tractors for petrochemicals. And these things too have taken them from themselves in space and in time.

And what happens to all this production, from the land and from the factories? Where does all the GNP go — except to faceless foreign exploiters in another country and a handful of rich in their own? And who the agents but their own rulers? In sum, what capitalist development has meant to the masses of these countries is production without purpose, except to stay alive; massive immiseration accompanied by a wholesale attack on the values, relationships, gods that made such immiseration bearable; rulers who rule not for their own people but for someone else — a development that makes no sense, has no bearing on their lives, is disorganic.

To state it at another level. The economic development that capital has super-imposed on the peripheries has been unaccompanied by capitalist culture or capitalist democracy. Whereas, in the centre, the different aspects of capitalism (economic, cultural, political) have evolved gradually, organically, out of the centre's own history, in the periphery the capitalist mode of production has been grafted on to the existing cultural and political order. Peripheral capitalism is not an organised body of connected, interdependent parts sharing a common life — it is not an organism. What these countries exhibit, therefore, is not just 'distorted' or 'disarticulated' development (Samir Amin), but disorganic development: an economic system at odds with the cultural and political institutions of the people it exploits. The economic system, that is, is not mediated by culture or legitimated by politics, as in the centre. The base and the superstructure do not complement and reinforce each other. They are in fundamental conflict — and exploitation is naked, crude, unmediated — although softened by artefacts of capitalist culture and capitalist homilies on human rights. And that contradiction because of capitalist penetration, runs right through the various modes of production comprising the social formation. At some point, therefore, the political system has to be extrapolated from the superstructure and made to serve as a cohesive — and coercive — force to maintain the economic order of things. The contradiction between superstructure and base now resolves into one between the political regime and the people, with culture as the expression of their resistance. And it is cultural resistance which, in Cabral's magnificent phrase, takes on 'new forms (political, economic, armed) in order fully to contest foreign domination'.[38]

But culture in the periphery is not equally developed in all sectors of society. It differs as between the different modes of production but, again as Cabral says, it does have 'a mass character'. Similarly at the economic level, the different exploitations in the different modes confuse the formal lines of class struggle but the common denominators of political oppression make for a mass movement. Hence the revolutions in these countries are not necessarily class, socialist, revolutions — they do not begin as such anyway. They are not even nationalist revolutions as we know them. They are mass movements with national and revolutionary components — sometimes religious, sometimes secular, often both, but always against the repressive political state and its imperial backers.

Notes

1. This is a development and reformulation of a paper originally given at the 'Three Worlds or One?' Conference, Berlin, June 1979.
2. GATT was set up to regulate trade between nations, the IMF to help nations adjust to free trade by providing balance-of-payments financial assistance, the World Bank to facilitate the movement of capital to war-torn Europe and aid to developing countries.
3. See 'Race, Class and the State', pp. 101-26, above, and 'From Immigration Control to "Induced Repatriation" ', pp. 131-40, above.
4. See *Ampo*, Special Issue, 'Free Trade Zones and Industrialisation of Asia', Vol. 8, no. 4 and Vol. 9 nos. 1-2, 1977
5. Even in the period of import-substitution — more succinctly described by the Japanese as 'export-substitution investment' — the multinational corporations were able to move in 'behind tariff barriers to produce locally what they had hitherto imported'. (*Ampo*, *op. cit.*)
6. The first Free Trade Zone was established at Shannon airport in Ireland in 1958 and was followed by Taiwan in 1965. In 1967 the United Nations Industrial Development Organisation was set up to promote industrialisation in developing countries and soon embarked on the internationalisation of Free Trade Zones into a global system. South Korea established a Free Trade Zone in 1970, the Philippines in 1972 and Malaysia in the same year. By 1974, Egypt, Gambia, Ivory Coast, Kenya, Senegal, Sri Lanka, Jamaica, Liberia, Syria, Trinidad and Tobago and Sudan were asking UNIDO to draw up plans for Free Trade Zones. (*Ampo*, *op. cit.*) Sri Lanka set up its Free Trade Zone last year soon after a right-wing government had taken power, albeit through the ballot box.
7. Jeff Frieden, 'The Trilateral Commission: economics and politics in the 1970s', *Monthly Review*, Vol. 29, no. 7, December 1977.
8. Richard Falk, 'A New Paradigm for International Legal Studies', *The Yale Law Journal*, Vol. 84, no. 5, April 1975.
9. Feroz Ahmed, 'Pakistan: the new dependence', *Race & Class*, Vol. XVIII, no. 1, Summer 1976.
10. *Ibid.*
11. *Dawn*, 13 May 1973, cited in *ibid.*

12. *Ibid.*
13. Cited in 'Oil for underdevelopment and discrimination: the case of Kuwait', *Monthly Review*, Vol. 30, no. 6, November 1978.
14. *Ibid.*
15. Selangor Graduates Society, *Plight of the Malaysian Workers in Singapore*, Kuala Lumpur, 1978.
16. Walter Easey, 'Notes on child labour in Hong Kong', *Race & Class*, Vol. XVIII, no. 4, Spring 1977.
17. Jon Stewart and John Markoff, 'The Microprocessor Revolution', *Pacific News Service*, Global Factory, Part II of VI.
18. Association of Scientific Technical and Managerial Staffs, *Technological Change and Collective Bargaining*, London, 1978.
19. D. Butler, abstract, *The Convergence of Technologies*, Report Series no. 5, Butler Cox Foundation, quoted in ASTMS, *op. cit.*
20. Ieuan Maddock, 'Beyond the Protestant ethic', *New Scientist*, 23 November, 1978.
21. *Ibid.*
22. See, for instance, I. Barron and R. Curnow, *The Future with Microelectronics*, London, 1979; C. Jenkins and B. Sherman, *The Collapse of Work*, London, 1979; Trades Union Congress, *Employment and Technology*, London, 1979; ASTMS, *op. cit.*; Colin Hines, *The 'Chips' are Down*, London, 1978; Chris Harman *Is a machine after your job?* London, 1979.
23. Of course there are those (guess who) who suggest that automation will not only release people from dirty, boring jobs and into more interesting work, but even enhance job prospects.
24. Cambridge Economic Policy Group, *Economic Policy Review*, March 1978.
25. Introductory address to the British Association for the Advancement of Science Symposium, *Automation Friend or Foe?*
26. Trades Union Congress, *Employment and Technology*, London 1979.
27. ASTMS, *op. cit.*
28. Harry Braverman, *Labor and Monopoly Capital*, New York, 1974.
29. Counter Information Services, *The New Technology*, Anti-Report no. 23, London, 1979.
30. Philip Virgo, *Cashing in on the Chips*, London, 1979.
31. Cited in Rachael Grossman, 'Women's place in the integrated circuit', *Southeast Asia Chronicle*, no. 66, Jan-Feb 1979.
32. *Ibid.*
33. 'Workers who must dip components in acids and rub them with solvents frequently experience serious burns, dizziness, nausea, sometimes even losing their fingers in accidents... It will be 10 or 15 years before the possible carcinogenic effects begin to show up in

women who work with them now.' *Southeast Asia Chronicle*, no. 66, Jan-Feb 1979.
34. Cited in *ibid.*
35. See A. Lin Neumann, ' "Hospitality Girls" in the Philippines', *Southeast Asia Chronicle*, no. 66, Jan-Feb 1979.
36. See 'The Day After tomorrow', by Peter Large, *Guardian*, 17 February 1979.
37. For the purposes of the general analysis presented here, I make no distinction between periphery and developing periphery.
38. Amilcar Cabral, *Return to the Source*, New York, 1973.

Race, Class and Caste in South Africa — An Open Letter to No Sizwe*

Dear Comrade,

I read your book on the national question in South Africa — *One Azania, One Nation*[1] — with great interest. But I have certain grave misgivings about your analysis of the elements in the theory of nation (chapter 6). Your discussion of course centres around Azania, but the questions you raise are not unrelated to the problems of other Third World countries. And it is for that reason — and in a spirit of enquiry and friendly discourse — that I take issue with you.

The crux of the matter is your discussion of colour-caste — the implications of that analysis for revolutionary practice. But since you try to clear the ground of 'terminological' and 'conceptual' obstacles before proceeding to your central thesis, it is to these issues I would first like to address myself.

1. You seem to be saying that to accept the concept of race — however used (anthropologically, biologically or sociologically) — is to accept a racial classification of people, giving each (race) a weightage or, in the alternative, denying it weightage (and therefore a hierarchy of superiority) altogether. 'For just as the supposed inferiority or superiority of races necessarily

* *Race & Class*, vol. XX, no. 3, Winter 1981.

assumes the existence of groups of human beings called "races", so does the assertion that "races" are equal in their potential for development and the acquisition of skill.' So that, for you, it is as meaningless to say that some races are superior to others as it is to say that all races are equal. Hence there is no such thing as race.

But you cannot do away with racism by rejecting the concept of race.

2. You deny the reality of race as a biological entity. Hence you deny the existence of racial groups. For the limited purposes of genetic science, however, you are prepared to describe such groups as 'breeding populations' — since 'such a description has no coherent political, economical or ideological significance'. But however you describe them — and however 'inherently' neutral the description — some 'breeding populations' do think of themselves as superior to other 'breeding populations' and act out that belief to their own social, economic and political advantage. Changing the description does not change the practice — but the practice can taint the description till that ceases to be neutral (so that for racism we merely substitute 'breeding populationism').

In the final analysis, it is the practice that defines terminology, not terminology the practice. The meaning of a word is *not* 'the action it produces' — as you seem to maintain with I.A. Richards. If so, to destroy the word would be to destroy the act — and that is metaphysics. On the contrary, it is action which gives meaning to a word — it is in the act that the word is made flesh. In the beginning was the act, not the word. Thus 'black', which the practice of racism defined as a pejorative term, ceases to be pejorative when that practice is challenged. Black is as black does.

You cannot do away with racism by using a different terminology.

3. Similarly, the use of the term ethnicity to differentiate between human groups that 'for some natural, social or cultural reason come to constitute a (temporary) breeding population' is equally irrelevant. In fact, it is, as you say, 'dangerously misleading'. For, in trying to remove the idea of group superiority while keeping the idea of group difference, ethnicity

sidles into a culturalism which predicates separate but equal development, apartheid. It substitutes the vertical division of ethnicity for the horizontal division of class, political pluralism for class conflict, and freezes the class struggle.

4. The concept of national groups implies, as you say, 'a fragmentation of the population into potentially or actually antagonistic groupings', and thereby facilitates 'the maintenance of hegemony by the ruling classes'. And the concept of national minorities, I agree, is essentially a European one and one that once again obscures the essential class nature of society.

5. But 'race' in its original sense of 'a group of persons or animals or plants connected by common descent or origin' (*Shorter Oxford Dictionary*) is no less neutral a term than 'breeding populations'. And that there are differences between such groups is an observable fact. What is *material*, however, is neither the term nor the group differences it implies, but the differential power exercised by some groups over others by virtue of, and on the basis of, these differences — which in turn engenders the belief that such differences are material. What gives race a bad name, in other words, is not the racial differences it implies or even the racial prejudice which springs from these differences, but the racist ideology that grades these differences in a hierarchy of power — in order to rationalise and justify exploitation. And in that sense it belongs to the period of capitalism.

6. Your 'central thesis', however, is that 'colour-caste' best describes 'the officially classified population registration groups in South Africa' — and that it is of 'pivotal political importance to characterise them as such'. About the importance of correct analysis for correct political action I have no disagreement. But, for that very reason, I find your characterisation of South Africa's racial groups as colour-castes not only wrong, but misleading.

Your argument for using the caste concept is made on the basis that South Africa's racial system (my phrase) has the same characteristics as the caste system in India. These refer to such things as rituals, privileges, mode of life and group cohesion ('an integrative as opposed to a separatist dynamic'). And whether or not 'the origin of the caste system in India is related to the question of colour', the crucial difference is that in India it is 'legitimised by cultural-religious criteria', whereas in South Africa it is 'legitimised by so-called "racial" criteria'. But in both, caste relations are 'social relations based on private property carried over in amended form from the pre-capitalist colonial situation to the present capitalist period'. To 'distinguish it in its historical specificity', however, you would characterise the caste system in South Africa as a colour-caste system — in which 'the castes articulate with the fundamental class structure of the social formations.'

But, in the first place, these are analogies at the level of the superstructure, with a passing consideration for the 'historical specificity' that distinguishes the two systems. They relate to ideological, cultural characteristics adjusted to take in considerations of class and social formations, but they do not spring from an analysis of the specific social formations themselves — they are not historically specific. That specificity has to be sought not in this or that set of religious or racial criteria, but in the social formation and therefore the historical epoch from which those criteria spring. The social formation in which the Indian caste system prevailed is qualitatively different to the social formation in South Africa, and indeed to that of India today. Secondly, it is not enough to say that caste relations are 'social relations in private property carried over in amended form' from a pre-capitalist era to a capitalist one, without specifying at the same time that private property in the earlier period referred mainly to land, whereas in capitalist society it refers also to machinery, factories, equipment. And that alters the nature of their respective social relations fundamentally. Thirdly, and most importantly, you make no reference to the *function* of caste. Caste relations in India grew organically out of caste functions of labour. They were relations of production predicated by the level of the productive forces but determined by Hindu ideology and polity. But as the productive forces rose and the relations of production changed accordingly, caste lost its original *function* — and, un-needed by capital, it was outlawed by the state. But because India, unlike South Africa, is a society with thousand modes, caste still performs some function in the interstices of its pre-capitalist formation and caste relations in its culture. South Africa, however, has caste relations without ever having had a caste function. Such relations have not grown out of a pre-capitalist mode; nor are they relations of production stemming from the capitalist mode. They are, instead, social relations enforced by the state to demarcate racial groups with a view of differential exploitation within a capitalist system.

To put it differently, caste as an instrument of exploitation belongs to an earlier social formation — what Samir Amin calls the tributary mode — where the religio-political factor and not the economic was dominant and hence determined social relations. The Hindu religion, like all pre-capitalist religions, encompassed all aspects of human life and Hindu ideology determined the social relations from which the class-state could extract the maximum surplus: the caste system. It is in that sense that India's great marxist scholar Kosambi defines caste as 'class at a lower level of the productive forces'. [2]

In the capitalist system, however, it is the economic factor which is dominant; it is that which determines social relations and, in the final analysis, the political and ideological superstructures. And how these are shaped and modified depends on how the economic system is made to yield maximum surplus value with minimum social dislocation and political discontent. Exploitation, in other words, is mediated through the state which ostensibly represents the interests of all classes.

Since European capitalism emerged with the conquest of the non-white world, the exploitation of the peoples of these countries found justification in theories of white superiority. Such attitudes were already present in Catholicism, but, muted by the belief that the heathen could be saved, found no ideological justification in scripture. The forces that unleashed the bourgeois revolution, however, were also the forces that swept aside the religious inhibitions that stood in the way of the new class and installed instead a new set of beliefs that sought virtue in profit and profit in exploitation. 'Material success was at once the sign and reward of ethical superiority' and riches were rather 'the portion of the Godly than of the wicked'[3] — and each man's station in life was fixed by heavenly design and unalterable. You were rich because you were good, you were good because you were rich — and poverty was what the poor had brought upon themselves. But to fulfil one's 'calling' was virtue enough.

In such a scheme of things, the bourgeoisie were the elect of God, the working class destined to labour and the children of Ham condemned to eternal servitude — 'a servant of servants... unto his brethren'. Each man was locked into his class and his race, with the whites on top and the blacks below. And between the two there could be no social mixing, for that would be to disrupt the race-class base on which exploitation was founded. To prevent such mixing, however, recourse was had to Old Testament notions of purity and pollution. Social or caste barriers, in other words, were not erected to preserve racial purity; rather, racial purity was 'erected' to preserve social, and therefore economic, barriers. The reasons for the racial divide, that is, were economic, but the form their expression took was social.

Thus, the racism of early capitalism was set in caste-like features — not ordained by religion, as in Hinduism, but inspired by it, not determining the extraction of surplus but responding to it. The Calvinist diaspora, 'the seed-bed of capitalistic economy',[4] would sow too the seeds of racism, but how they took root and grew would depend on the ground on which they fell.

In general, however, as capitalism advanced and became more 'secular', racism began to lose its religious premise and, with it, its caste features and sought validity instead in 'scientific' thought and reason — reaching its nineteenth-century apogee in Eugenics and Social Darwinism. Not fortuitously, this was also the period of colonial-capitalist expansion. But at the same time, with every advance in the level of the productive forces and, therefore, in the capitalist mode — from mercantile to industrial to finance and monopoly capital — racist ideology was modified to accord with the economic imperative. Slavery is abolished when wage-labour (and slave rebellion) makes it uneconomical; racism in the colonies becomes outmoded with the advent of neo-colonialism and is consigned to the metropole with the importation of colonial labour. And within the metropoles themselves, the contours and content of racism are changed and modified to accommodate the economic demands (class) and political resistance (race) of black people. Racialism may yet remain as a cultural artefact of an earlier epoch, but racism recedes in order that capital might survive.[5]

But not in South Africa. There, though the economy is based in the capitalist mode, the superstructure bears no organic relationship to it. It does not on the whole respond to the economic imperatives of the system. And that inflexibility in turn inhibits the base, holds it down, prevents it from pursuing its own dynamic. Hence, there is a basic contradiction between the superstructure and the base.

Where the contradiction is located, however, is in that part of the superstructure which relates to the black working class — and black people generally. In effect, there are two superstructures (to the same economic base) — one for the whites and another for the blacks. The white superstructure, so to speak, accords with the economic imperatives — and is modified with changes in the level of the productive forces and of class struggle. It exhibits all the trappings of capitalist democracy (including a labour movement that represents the interests of the white working class) and of capitalist culture (except when it comes to mixing with the blacks). For the blacks, however, there is no franchise, no representation, no rights, no liberties, no economic or social mobility, no labour movement that cannot be put down with the awesome power of the state — no nothing. The 'black superstructure' in other words, is at odds with the capitalist economy, sets the economy at odds with itself, and inhibits its free development — so that only changes in that superstructure, in racism, can release the economy into its own dynamic. South Africa, therefore, is an exceptional capitalist social formation.

In the second place, South Africa's racist ideology, compared to that of other capitalist societies, has not changed over the years. Instead, it has gathered to itself the traits, features, beliefs, superstitions, habits and customs of both pre-capitalist and capitalist social formations. Its caste features bear an uncanny resemblance to the Hindu caste system of medieval India, though we know them to be inspired by Calvinism, the religion of capital. It combines, in Ken Jordaan's exact phrase, 'the Afrikaners' fundamentalist racialism with the instrumentalist racism of British imperialism'.[6] It finds authority in religion and in science both at once — in the doctrines of the Dutch Reformed Church and the teachings of Darwin. ('At the birth of the Union of South Africa,' says Jordaan, 'Calvin and Darwin shook hands over the chained body of the black.') It is enforced by a capitalist state and receives its sanction from the church. And it is as open, obtrusive and unashamed as the racism that once justified the trade in human beings.

7. But what are the material conditions that made South Africa's racist ideology so intractable? What is the significance of the modifications that are currently being made in the racist structure?

These are not your questions and I am not competent to answer them, but you (and Johnstone[7]) go some way to answering the first in implying that South African capitalism was neither colonial nor industrial (in the strict sense), but extractive — derived from diamond and gold mining. Which meant that the labour process called for a mass of unskilled labour which was found in the native black population — and a docile workforce which could be fashioned by racism. Hence, the nature of early South African capitalism reinforced and did not loosen up on racism as, for instance, in the USA.

Secondly, South Africa was a settler society which neither assimilated itself into the indigenous social structure (as Aryan India) nor was able to decimate the native population (as in the USA or the Caribbean). The settlers instead were (and are) a slender minority, distinguished by race and colour, faced with a massive black population. (The only parallel is Zimbabwe.) Hence, the only way they could preserve their economic privileges and their political power was to stand full-square against the encroachments of the black masses.

But — and here I am addressing myself to the second question — the demands in the economic imperative, both nationally and internationally, can no longer be ignored. Hence, Botha's attempts to 'modernise' racism — to accord with monopoly capital — by removing its caste features.

There are other changes, however, which have been in train for a longer time — and which are more dangerous. And these, as you rightly point out, are the creation of a black comprador class (comparatively negligible) and, more importantly, of black 'nations'.

In theoretical terms, what these strategies hope to resolve is the contradiction between superstructure and base, and so release the economic forces without incurring the loss of (white) political power. First, by removing the caste barriers and thereby providing social and economic mobility for the black working class within the central social formation. Secondly, by removing the superstructure for the blacks into a social formation of their own, a black state, in which they would appear to govern themselves while still being governed. The conflict, in other words, is extrapolated into a different (black) social formation — which is then subsumed to the needs of the central social formation, thereby maintaining, as you say, the hegemony of the ruling classes — to me, the white ruling class. For, surrounded as South Africa is by black African nations — and given the lesson of Zimbabwe — there is no way it is going to cede an iota of white power of which racism is the guarantor. So that even if, at some far point in the future, racism dies for capital to survive, it will have to be resurrected — for capital to survive in white hands.

8. Which brings me to my final point. You say — and perhaps you are forced into saying it by virtue of your colour-caste interpretation (and on behalf of marxist orthodoxy) — that in the final analysis, the struggle in South Africa is a class struggle, to be waged by the working class as a whole, black and white alike.

But, as I hope I have shown, South Africa is the one capitalist country (Zimbabwe might have gone the same way but for black guerrilla struggle) where ideology and not production relations determines white working-class consciousness.[8] That is not to say that there are no class contradictions between white capital and the white working class, but to say that — vis à vis the black working class — the horizontal division of class assumes the vertical division of race: the horizontal is the vertical. Class is race, race class. In other words, so long as the blacks are forced to remain a race apart, the white working class can never become a class for itself. And as for the blacks, if the unending rebellion of the past few years and the birth of the Black Consciousness Movement are anything to go by, they are fast becoming both a race and a class for themselves — and that is a formidable warhead of liberation. In sum, the racist ideology of South Africa is an explicit, systematic, holistic ideology of racial superiority — so explicit that it makes clear that the white working class can only maintain its standard of living on the basis of a black under-class, so systematic as to guarantee that the white working class will continue to remain a race for itself,[9] so holistic as to ensure that the colour line is the power line is the poverty line.

To reiterate, in its ability to influence the economic structure — rather than be influenced by it — South Africa's racist ideology belongs to a pre-capitalist social formation but, anachronistically, is present in a capitalist one — thereby distorting it. (In other words, it is not a pure capitalist social formation.) The emphasis on the ideological instance produces a characterisation of the population groups in South Africa as caste groups demarcated on colour lines (and 'articulating' with the class structure); an emphasis on the economic mode produces a straightforward (marxist) race-class concept and characterisation — thereby leading one to conclude that the economy in trying to burst its bonds would burst also the racist nexus. But if they are both comprehended equally and at once, holistically, South Africa shows up as an exceptional capitalist social formation in which race is class and class race — and the race struggle is the class struggle.

Notes

1. No Sizwe, *One Azania, One Nation: The National Question in South Africa*, London, Zed Press, 1979.
2. D.D. Kosambi, *The Culture and Civilisation of Ancient India*, London, 1965.
3. R.H. Tawney, *Religion and the Rise of Capitalism*, London, 1975.
4. Gothein quoted in Max Weber, *The Protestant Ethic and the Spirit of Capitalism*, London, 1965.
5. Racialism refers to attitudes, behaviour, 'race relations'; racism is the systematisation of these into an explicit ideology of racial superiority and their institutionalisation in the state apparatus.
6. K. Jordaan, 'Iberian and Anglo-Saxon racism', in *Race & Class*, Vol. XX, no. 4, Spring, 1979.
7. F.A. Johnstone, *Class, Race and Gold*, London, 1976.

8. This may be a heresy, but South Africa is a country that invites heresies.
9. Note, for instance, how white workers have recently demonstrated their unrelenting opposition to blacks moving up into skilled jobs — thereby serving to entrench white racial superiority and engendering fascist attitudes, which the state could well exploit.